PRAIRIE FARMING IN AMERICA.

WITH NOTES BY THE WAY

ON

CANADA AND THE UNITED STATES.

BY

JAMES CAIRD, M.P.

AUTHOR OF "ENGLISH AGRICULTURE," "LETTERS ON THE CORN CROPS," "HIGH FARMING," "THE WEST OF IRELAND," ETC.

NEW YORK:
D. APPLETON AND COMPANY,
346 & 348 BROADWAY.
1859.

CONTENTS.

INTRODUCTION.

LETTER I.

LETTER II.

LETTER III.

LETTER IV.

LETTER V.

LETTER VI.

LETTER VII.

LETTER VIII.

LETTER IX.

LETTER X.

LETTER XI.

PRAIRIE FARMING IN AMERICA.

INTRODUCTION.

Present Position of Agriculturists in England.—Continued Rise of Rent.—Increased Rate of Wages not equally certain.—Competition in the Hire of Land.—Whence it arises.—Foreign Produce.—That of Lands of superior Fertility chiefly affects us.—Increasing Value of Live Stock.—Home Competition the true Source of diminished Profits.—Extent to which the Business of Farming is giving place to other Occupations.—Area of ordinary Farming thus curtailed.—Necessity for Farmers to thin the ranks of Home Competition.—Daily Consumption of Foreign Grain.—Offers a Good Prospect to the Emigrant.—The Prairies of Illinois a suitable Field for him.—Their Advantages.—Price.—Ague.—Stock Farming and Indian Corn.—Cost of Voyage and Journey.—The present a very favourable Time.

The present position of the agricultural body in the United Kingdom is interesting and peculiar. The landowner and the agricultural laborer are both profiting by the same cause, a limited supply of the commodity in which they deal. So long as this country continues to prosper, the value of land must increase, for there can be no increase of the land itself. But the demand for labour varies, and the supply is subject to causes which render it uncertain. While, so long as the present system of taxation continues, there must be a continued rise in the value of land, there appears to me no equal certainty of a progressive advance in the rate of wages.

But the hirer of land, the farmer, must inevitably suffer

from the continued competition for its possession. He has not only to meet his own class, a necessarily increasing body, in this competition, but to contend with men who, having made money in other pursuits, wish to retire to the more pleasurable occupations of a country life. It is this competition which is the true cause of the reduced profits of farming, and this is more likely to increase than diminish. Great Britain is the most attractive place of residence on the surface of the globe, whether we regard its equable and healthy climate, its varied scenery and field sports, the almost sacred character of the rights of property in the eyes of its people, and the admirable combination of liberty and order which is preserved under its political constitution. Men will pay for these advantages, when they can afford it, a price which is not measured by the ordinary rates of profit.

Besides this competition, which raises the rate of rent, the farmer must now meet in his own market the produce of lower-priced foreign lands. He will, no doubt, always have the cost of transport in his favour, and this would generally be sufficient to balance the difference of rent; but the land of this country cannot be cultivated without manure, and the farmers of those foreign countries whose soil is rich enough to yield corn for many years without manure, are thereby able to undersell the British producer in his own market. The cost of labour when the value of food of the working stock is calculated, is nearly the same at home and abroad, and superior fertility alone will be found to turn the advantage in favour of the foreign producer.

The special adaptation of Britain for the production of live stock, and the constantly increasing demand for that branch of the farmer's produce, have hitherto modified the effects of foreign competition in corn. But even these, excellent though they have proved, cannot permanently counteract the cause of the farmer's diminished profits: viz., home competition for the possession of land. The soil here is now becoming more valuable for other purposes than ordinary farming, and the proportion between the producers and consumers of food is undergoing a rapid change. It appears from the Census that, in 1851, only 16 per cent. of the adult population of England was occupied in the business of agriculture. During the previous twenty years the proportion had fallen from 28 to 16 per cent., from no actual decrease of the numbers employed in agriculture, but from the far greater proportional increase of trade. The same gradual change is going on. At this time there is probably not more than one-tenth of the adult population of England employed in the culture of the land. The manufacturing, mining, and town populations are thus gradually absorbing the business of the country, increasing the value of the land and the profits of the landowner, but in the same proportion diminishing the area left for ordinary farming.

The time seems thus to have arrived when the farmers must thin the ranks of home competition by sending off the young and enterprising to countries where they may become the owners of a fertile soil, and profitably contribute to supply the wants of the old country, whose land can no longer meet the demands of her dense population. During the last year we have imported into this country at the rate of nearly one million quarters of grain each month. We have thus in addition to our home crop, consumed each day the produce of TEN THOUSAND acres of foreign land, a demand so vast as to offer to young men of our own country the strongest inducements to take their share in its supply.

Having, during last autumn, had an opportunity of making a pretty careful inspection of a part of the valley of the Upper Mississippi, probably the most fertile corn region in the world,

I have collected for publication, in the form of a series of letters, the notes made by me at the time. There may be other countries which present equally good prospects to the agricultural emigrant. I venture to speak only of that which I have seen. This seems to me to offer the very field which we want at present,—a virgin soil of easy culture, with no forests to clear, of extraordinary natural fertility,* in a country traversed by a most perfect system of railways, where no settler need be more than ten miles from a station, whose shore is washed by one of those great lakes through which an outlet is found to the Atlantic, and which possesses in the Mississippi itself a vast artery of commerce, navigable by steamers for thousands of miles. A great part of the country is underlaid with coal, iron, and lime, thus affording a present supply of such minerals, and the prospect of a great increase of value should the people ever turn their attention to manufactures. There is a complete organisation of markets throughout the country; and, setting aside the export to England, there is a very large and increasing local demand for every article of agricultural produce. The price of labour is economised by the most extensive and profitable use of agricultural machinery, and by the comparatively small cost of maintaining horses and working cattle. The grazing of cattle and sheep is very profitable, and the production of merino wool, already large, admits of vast increase.

The fee simple of this land can be purchased at from 40*s.* to 50*s.* and 60*s.* an acre.

As a mere investment, this land would pay well to purchase and hold for a few years, and the increasing supply of gold, of which America herself yields an annual crop of ten millions sterling, will every year contribute to the higher rel-

* See "Chemical Analysis of Soils" by Professor Voelcker, Appendix I.

ative value of land here and elsewhere. But the British emigrant, when he purchases this land, secures to himself not only the profits of farming it, but has also the growing increase in the value of the land itself, a right to which he can have no share at home. The country is now brought within a fortnight's journey from our shores, and is actually more accessible from Great Britain than most parts of Ireland were fifty years ago.

There is one drawback to which I have several times adverted in these letters. Besides the ordinary hardships to which men are content to submit when they leave the comforts of an old country for the prospect of future great benefit to themselves and their families, emigrants to rich new countries, south of the 45th degree of latitude, are in certain seasons more or less liable to an attack of ague. It is very fatal to old people, but the young seldom suffer more than temporary inconvenience from it; and the climate which produces ague is nearly free from some other and more fatal complaints which are met with in colder latitudes. By degrees the emigrant becomes acclimated, and very many never experience the disease at all. Care and remedial measures prevent or remove it, and it gradually disappears with the general settlement and cultivation of the country. The population returns prove that the ague has no serious effect on the health of the people. A country whose people double in ten years, as Illinois did between 1845 and 1855, cannot be very unhealthy. Indeed, compared with many States in the Union, Illinois stands high in the tables of health. She is before New York, Connecticut, and Massachusetts, Eastern States which are generally deemed healthy. And in comparison with England, the mortality in Illinois contrasts most favourably,—De Bow's compendium of the seventh census of the United States showing that her death rate in 1850 was less than 14 in the 1000, while that of Eng land is at present rather more than 24. While I think it right to direct the emigrant's attention to the ague as an element in his calculations, it forms in reality a very small counterpoise to the many advantages which are open to those who make judicious and well-chosen settlements on the Western Prairies.

There are two branches of his business to which I would specially ask the attention of the British emigrant to Illinois, viz. stock farming, and the cultivation of Indian corn. Full details will be found on both subjects in these letters. A good stock of cattle or sheep can be bought by a comparatively small outlay of capital; and, so long as the open Prairie is thinly settled, grass for half the year may be had for nothing, and hay for the other half for only the cost of saving it. In regard to Indian corn, both climate and soil are more suitable to it than wheat. It can be grown to any extent, with a certain measure of success, every year, and, unlike wheat, this grain may be harvested with safety over a period of many weeks. A small and regular supply of labour thus suffices for the management of a large extent of land. There is always a market for it, and the lowest price at which we have ever seen it in England will afford a very good return to the Prairie farmer of Illinois, after deducting all the charges of transport.

An emigrant from this country may be set down in Illinois at a total cost from Liverpool or Glasgow of 6*l.* 7*s.*, inclusive of provisions.

The present is a most favourable time for commencing to farm in Illinois. The panic of 1857 has not yet been forgotten, and the prices at which every sort of contract (building, fencing, ploughing) may be executed, are 50 per cent. below the average rates.

LETTER I.

Voyage in the "Persia."—New York.—The Hudson River.—Barren Soil.—American Railways.—Saratoga.—Extravagant Prices.—Lake George.—Lake Champlain.—Sunshine and Storm.—Burlington in Vermont.—Maine Liquor Law.—The Boundary Line.—Poor French Canadians.—The St. Lawrence.—Montreal.—The Victoria Tubular Bridge.—Grand Trunk Railway.—Ottawa.—The Lumber Trade.—Ottawa or Montreal as the Capital.—Shorter Water Route to the Western Lakes.—Value of Land.—Grants of Land.—Prescott.—Kingston.—Coburg.—Process of clearing the Forest.—Toronto.—Hamilton.—Complaint of low Wages and want of Employment.

On the 4th of September, 1858, I embarked with a friend on board the "Persia" at Liverpool, and loosed from our moorings in the afternoon to proceed on our voyage to New York. The ship had a full cargo and more than 200 cabin passengers. Great order and regularity prevailed on board, and though we had heavy head winds all the way across the Atlantic, and two severe gales, the voyage on the whole was pleasant and prosperous. On the evening of the twelfth day we reached New York.

The bright clear sky and the sunny look of the houses and public buildings, with the frequent cafés, reminded us that we were now in the latitude of Naples. And the appearance of the people was so different from that of Englishmen that we almost felt surprised to hear them speaking the English language. Everything was new and pleasant, except the manners of the people, and the extortionate charges of every one from whom it was necessary to obtain the slightest service.

After spending a short time pleasantly in New York we left it for Montreal, taking the Hudson River route to Albany. I was somewhat disappointed with the far-famed scenery of this river, the banks of which (as indeed is the case with land everywhere in America at this season) are strikingly deficient in verdure. The aspect of the rocky cliffs and the broad river is very pleasing, enlivened as the latter was by numerous white-sailed little sloops passing up and down, and now and then by a white steamer trailing half a dozen loaded canal boats. After passing West Point the wooded hills in the foreground are backed by the Catskill Mountains, and the reaches of the river are here extremely picturesque. But the wood, though there is plenty of it, is little better than copse or brushwood. There are no fine trees, and the soil appears to be beyond measure barren. There are houses here and there, peeping out from the woods, with green outside blinds, and towers with tall steeples, all very white. But in the 150 miles between New York and Albany I saw no good land and no grass which any British ox would touch. From Albany to Troy the railroad cars were crammed with people, all hurrying about, and yet it was difficult to see what occupied them, as there are few signs of manufacture, and the soil is evidently unfruitful and neglected. There seemed nothing here to attract or create much capital. No cheapness of price would compensate for the natural inferiority of such land to good land in our own country. The lowland along the river is marshy and aguish-looking, and the upland seemed either bare rock or stiff clay. Between Albany and Troy there were some patches of Indian corn, but the country is very uninviting,—no verdure, and no pleasant homesteads.

I liked New York from its variety and picturesque novelty. But the country so far disappoints me, and the same barren aspect, changed from clay to blowing sand, continues to Lake Champlain. The railways are very uneven and uneasy. When I awoke after my first day's ride I thought myself at sea again, feeling the sensation which one experiences after a stormy voyage. This was at Saratoga, the great watering-place to which the Americans resort during the fashionable season. The season was now over, and only a few lingering visitors remained. The town is an assemblage of hotels, at which you are "taken in and done for" at 10*s.* 6*d.* a day. The hotel in which we stayed is the largest. It has beds for about 1000, and the waiters are all black, slow, and not very obliging. The prices of everything are enormous. A guide-book costs 8*s.* 4*d.*, a bottle of sherry 8*s.* 4*d.* to 10*s.* 6*d.*, Madeira 24*s.*, Bass' beer 2*s.* 6*d.*, a cab for three miles 8*s.* to 10*s.* The horses are not unlike their masters. They hold up their heads, shake their little cocktails, and away they fly with the spider-wheel carriage which the fast Yankees drive. They seem in a desperate hurry at the start, and yet I have not found them more enduring than an English horse at the end of the day's journey.

Lake George may be visited by spending another day, and it will well repay the time, being exceedingly beautiful, more so than Loch Lomond, which it resembles in size. The mountains are not so high, but are wooded to their summits. The hotel accommodation is excellent, and good sport may be had by boat-fishing. In driving down from Lake George to Ticonderoga, you pass through a little village with four conspicuous churches. One of them looks neglected; the driver "guesses she was a Baptist, but he reckons she didn't pay, and they stopped running her."

At Ticonderoga we embark in the steamer up Lake Champlain, the day so extremely hot and calm that the most shady and airy place is welcomed. For some hours the fast steamer speeds on at fifteen miles an hour through the smooth water, the scene on the left presenting a pretty foreground of farms along the lake, backed by the lofty wooded mountain range of this part of the State of New York. On the right the land is broken into rounded hills, with patches of woodland and corn, but chiefly in grass, green for this country, and giving to the State its character and name of Vermont. Suddenly we feel a whiff in our faces. The wind has changed, a dense blackness is covering the sky northwards, and in half an hour we have sailed into a thunderstorm. It was accompanied and followed by a gale, and in another half hour the lately placid summer lake was surging like the Atlantic. Our white gala-day steamer was quite unfit to cope with this, and the shakes she gave when struck by a heavy sea shook the nerves of most of the passengers on board. By the evening we reached Burlington, on the Vermont shore, where, with other Englishmen, we landed for the night, at a large hotel, with the hope of a good supper after our storm-tossing. But we had reckoned without our host;—we found ourselves in a temperance State, actually under the Maine Liquor Law,—plenty to eat, but not a drop of anything stronger than water to drink, without a medical certificate. The men looked no better than those in the other States, the bar room being filled with idlers chewing and spitting. The women certainly seemed more fresh and ruddy, yet it is odd that they alone should show the good effects of a law which could not have been intended for the gentler sex.

Next morning we proceeded by railway towards Montreal, crossing Lake Champlain at a narrow point by a long wooden bridge, and shortly afterwards entered Canada. The line between the two countries is an ideal one, not more definite than the march of two Highland farms. The land is cold and poor, held in strips by French Canadians, whose listless gait and lean cattle betoken a poor business. The notices at the railway crossings are in French, and the country people at the stations, and those travelling by the railway, converse in French. The weather was cold, the thermometer having fallen since the previous morning from 74 to 44.

The country is very flat, but the land improves in quality as we approach the St. Lawrence, which, opposite Montreal, is a magnificent stream, like an arm of the sea, rushing with a great tide, two miles broad. From the ferry steamer the city is seen to great advantage, its wharves stretching along the river, and the tin roofed houses and church cupolas sparkling brilliantly in the sun. It is a remarkably handsome town, backed by a lofty wooded hill, *the* Mountain, which all strangers are expected to visit, and from which the prospect is very extensive. One striking object is the Great Victoria Tubular Bridge now being constructed across the St. Lawrence, two miles in length, as the viaduct of the Grand Trunk Railway, a prodigious engineering work, but which, when completed, will in all seasons, summer and winter, afford to the railway a continuous outlet to the Atlantic for the whole stream of traffic from British America and the North Western States of the Union. The vast expenditure on this railway and its works has greatly enriched Montreal.

From Montreal I proceeded to Ottawa, taking railway to La Chine, and there embarking on a wide lake-like water which is the confluence of the rivers St. Lawrence and Ottawa. Some miles farther up we reach the rapids of St. Anne, the scene of Moore's song, and the point at which the Grand Trunk railway crosses from the island of Montreal to the mainland by another large engineering work, a tube and bridge, five hundred yards long. Beyond this the stream divides, sending part of its waters along the northern side of the island of Montreal, which is nearly forty miles in length and fertile, farmed by French peasants and English and Scotch farmers. On the opposite northern shore a fine tract of wooded country, about ten miles square, has been reserved for an Indian tribe, whose village is on the shore. They are said to be completely controlled by French priests, who suffer no intercourse with settlers, and allow no wharf to be made in case it might encourage traffic.

We escape another series of rapids by changing to railway for ten miles, through a wooded country, which, where partially cleared, is nearly covered with huge boulders of granite, a poor country, supporting a poor French population. We then again embarked on the river. Its banks nearly all the rest of the way to Ottawa, sixty-five miles, are low and swampy, farther back partially cleared, with wooded highlands on the north shore in the background. At the outlets of the rivers falling into the Ottawa there are generally saw-mills and large piles of "lumber" or sawn timber, the Rideau Falls thus being made to cut up 170,000 logs in a season.

The city of Ottawa is finely situated on the summit of the bank. From the flagstaff on the hill the view of the falls, the river, and the surrounding country is extremely fine, and nothing can excel its position as the site of a town, strongly placed for defence if need be. But except with a view to a future plan, if such were contemplated, of a shorter way to Lake Huron from the Atlantic, one cannot imagine why a place like this, in the wilderness and out of the present line of traffic altogether, should have been chosen by the home government in preference to Montreal as the capital of Canada. Even if Ottawa should become the route to the West, Montreal is still the outlet to the ocean through which all must pass, and is already a rich and populous city, with all the necessary public buildings, barracks, and accommodation for the seat of government. It would be little more absurd to decree the removal of the seat of government from London to Norwich than it was to prefer Ottawa to Montreal.

There can be no doubt that sooner or later a shorter and better water route from the Atlantic to Lake Huron than that through the Welland Canal must be constructed. Not only may the route be vastly shortened, but what is of even more consequence, a canal with capacity for ships of 1000 tons is wanted. If that were provided the teeming harvests of the West could be shipped directly to the Atlantic, and the position of Canada enables her to command this traffic. The shortest route, if there is depth of water, would be by connecting the navigation of the Ottawa river with French river, down which a passage may be opened to Lake Huron. This would cut off 450 miles of the present water route between the shores of the three western lakes and the Atlantic. But if adequate depth of water is not to be had by that way, a ship canal between Toronto and Georgian Bay would probably secure it. The Canadian Government and Legislature would do a great service to their country by the early construction of a ship canal in the best route possible, not only to secure the transit of the Western States, but to serve the territory on the north shore of Lake Superior, and of all that vast country on the Red River, and the Saskatchewan, which will shortly be opened to their enterprise, and more especially if it should be determined to make this the route to the gold region of Fraser River and British Columbia.

I visited the Chaudiere Falls by moonlight. The view from the suspension bridge is extremely fine, the body of water magnificent, and the varied forms into which the water breaks, all along the extended face of the rocks, are very beautiful. This is the great seat of the lumber trade. The logs, cut far back in the woods of Canada, find their way down the river to this point, red pine, yellow, and white,—and a kind, the hardest of all, Norway pine. Here many great saw-mills are at work, about 200,000 logs being cut up in a season.

Farther up the river the land is said to be much better than that below Ottawa. Wheat is grown with some success, and for cleared farms large prices are asked. I was told of one farm, highly improved, about eighteen miles above Ottawa, for which as much as 20*l.* an acre was expected. This very greatly exceeded any estimate that I could form of the value of the land in such a locality. The summers are very hot, and the musquitoes abound in the woods to such an extent as to render it impossible almost to live in them at that season; and the winters are long and severe, costly in fodder for cattle over such an extended period.

Between the Ottawa river and Lake Huron free grants of land, not exceeding 100 acres each, are made to settlers, on condition that they build a log hut and reside on the lot, and that they clear and bring under cultivation twelve acres in four years. These twelve acres are probably reckoned the measure of what the labour of one man and his family may be reasonably expected to clear in a period of four years. The whole of this country is covered with wood: the soil being more sandy than most other parts of Lower Canada, it is found better suited for autumn wheat; but it is generally too poor, even when cleared, to be profitable.

From Ottawa to the St. Lawrence at Prescott the country presents a similar character. On the shore of the river there are some good farms tolerably cultivated, with hop-gardens on favourable spots. From this point to Kingston, and thence to Coburg, the country is but partially cleared; very often the train shoots for many miles together through the primeval forest, a path having been cut in the woods for the railway track, and the felled trees and branches still lying where thrown, on both sides of the line.

Northwards of the line, between Coburg and Toronto, there is a better tract of country, which has well repaid the labour of the farmer. Formidable though the cutting down of the forest appears, to the strong arm of a healthy young man, it is one of the most pleasing of out-of-door occupations. Though the winters are cold the weather is steady, and an active man may work in his shirt sleeves in the shelter of the forest during most of the clearing season. To cut down the trees on an acre of land, and pile them with the branches ready for burning, is reckoned a good month's work, and by steady perseverance, six acres may thus be accomplished during the winter. When the ground has become dry and warm, the piles are set on fire, the ashes are afterwards scattered over the surface, and early in autumn the ground is sown with wheat, which generally proves a good crop. On fair land, in a good situation, a hard-working man, if he gets the land for little or nothing, may soon earn a livelihood in this manner. But if he has to pay such prices as I heard quoted,—*5l.* to *6l.* an acre in any eligible locality,—and reckons the value of his own labour in clearing, and the loss of time during which he has to wait for his first crop, one can feel no surprise that the tide of emigration has of late years set steadily westward to the open prairies, where the land costs less to purchase, and from which a crop may be reaped in the first year of settlement.

Toronto is a fine city, with an excellent harbour on Lake Ontario: the harbour is protected by a low neck of land which forms a natural breakwater. Wide streets, numerous churches, and public buildings, with splendid stores and shops, betoken a place of growing prosperity. Toronto is the outlet of a good agricultural country, and, should a ship canal be made here to connect with Lake Huron by the Georgian Bay, the business of this flourishing city as a port of transit would be materially augmented. By means of the Grand Trunk and Great Western Railways it already possesses every facility for communication by land.

The drive towards Hamilton along the shore of the lake, which lay quite smooth and calm, was beautiful. The country is about half cleared, a heavy wheat soil, on which the new-sown wheat fields were all carefully water-furrowed. In some places the land was a stiff red clay, but most of it a heavy brown clay loam. There seemed to be very little Indian corn, and the grass on the pastures was either bad, or quite eaten off, or scorched up.

As the line nears Hamilton we pass by a wooden bridge over a chasm, which by the breaking of the bridge was the scene of a frightful railway accident some time ago. This town is placed on a bay at the head of Lake Ontario, from the waves of which the harbour is protected, like Toronto, by a natural breakwater. There are several handsome streets and houses in the town, and the villas in the neighbourhood are as substantial and elegant as those in the vicinity of our best towns in this country, with shrubberies, lawns, and greenhouses kept in the nicest order. From Hamilton to Niagara the railway runs along a broad tract of low country, stretching from a range of high tableland on the right to the shores of the Lake. At all points where we stopped for an excursion into the country, there was a uniform complaint among the Irish labourers of low wages and want of employment, and the wheat crop in this part of the country had proved a very short one.

LETTER II.

Falls of Niagara.—Canada West.—Mode of Farming.—Short Wheat Crop.—Average Produce.—London.—Price of Land.—Climate.—Diseases produced by Malaria.—Rich Lands more subject to them than poor.—Proposed Route to British Columbia.—Red River and the Valley of the Saskatchewan.—The Hudson Bay Territory.—Alleged Fertility of the Country.—Failure of the Selkirk Settlement.—Plague of Grasshoppers.—Mr. Kitson's Account of the Settlement on Red River.—Policy of Abandoning that Country to Canada.—Probable Over-estimate of its Value.

TOWARDS evening the train landed us at Niagara, but we caught no sign, either by sound or sight, of the great Falls till we found ourselves seemingly close in front of them at the Clifton House Hotel. The doors and windows of this hotel shake day and night, though it is really a mile distant from the Falls, and the sound seems no greater when you are close beside them than it is here. Following Sidney Smith's example at Woodhouselee, I pinned or wedged my door and window with complete success, then took a moonlight view of the Falls, during which we had the good fortune to see a lunar rainbow. After a two-mile walk to the suspension bridge, I regret to be obliged to confess that my first impression was one of disappointment. The country is tame and flat though wooded, and the river leaps from this flat *into* a deep gorge, on which you look down, instead of finding yourself in a valley *from* which you might look up. During the night you are roused by a sound like a fearful storm, but it continually changes, and presently you might imagine that you were close to 1000 railway engines blowing off their steam. How eagerly one springs up to get the first view in the morning sun!

We were fortunate in making the acquaintance of Mr. Church, the Turner of America, whose great picture of Niagara was not long ago exhibited in London. He was still here, studying and sketching this ever-changing scene. With him as our guide we soon discovered cause for wonder and admiration. Descending by a steep path to the bed of the river, we cross in a ferry-boat below the Falls, and are drawn up an inclined plane in a tunnel nearly 200 feet to the top, on the American side. Here, from the bent branch of a cedar literally hanging over the precipice, we obtained a view along the entire face both of the American and Horse Shoe Falls. We next crossed a little neck of land which brought us to the edge of the river above the Fall, and within a few feet of the brink,—the water here running so clear and shallow that a child might wade in it with safety. Some hundred yards further up we pass by a bridge to Goat Island, a picturesque spot, covered with natural wood some seventy acres in extent, which divides the Falls. Crossing this island we find ourselves in presence of the grand arm of the river, where the rushing waters, surging six to ten feet high, are pouring towards the Horse Shoe Fall. We pass by a frail wooden bridge to a tower built in the water, and close to the edge of the precipice. This we ascend, and there, close below and before us, rushing on now as it has done every day and night for 6000 years, was this tumultuous sea. Its chief grandeur is the central mass of green solid water, which glides unbroken over the Fall, ten thousand tons a minute, into the horrible abyss, 160 feet below. Noise there is enough, but it does not seem to come from that smooth tongue of water which in a moment would suck into destruction the Great Ship herself. A magnificent rainbow spanned a great part of the gorge below the Fall, and, for some hundred yards, the surface, clear of the spray which rises in front, is like one mass of churning cream. Recrossing Goat Island we had a second view of the American Fall, which has little of the grandeur of the Canadian side, the body of water being smaller and more shallow. There is here a little islet, connected with Goat Island by a slight foot-bridge to which there is attached a fearful story. A foolish man carrying his child across the bridge, held it by way of joke over the torrent; in its horror the child struggled and fell out of his arms. He sprang after it, and in one moment both were in the abyss.

Returning to the Canadian side of the river, I was conducted down a very long stair to the edge of the water, and thence by a narrow footpath under the Table rock to the foot of the Fall, behind which, on a footpath of shingle and rock, slanting outwards to the horrible cauldron, we stand and look through the Fall, which is pouring out above, but clear of us in front. We then follow the guide a little further. A very few yards bring us to a point of rock beyond which there is nothing but sheer precipice. There is said to be no danger, but I was almost smothered with the spray, faint for want of breath from the absolute loss of air in the thick flying foam. At this point it is hardly safe to turn with the face towards the Fall, and a very short stay here will satisfy the most curious. Getting back to a more secure position, we stood for some time watching the sun as now and then he was to be seen through a momentary thinness or opening in the falling water.

On reaching the top of the bank, we drove up the Canadian side to view the rapids, which are perhaps equal in grandeur to any part of the scene. Standing by the river side and looking up the stream, we see nothing whatever between the horizon and ourselves but the great river roaring and leaping down, miles broad apparently, though that is an illusion; but there it comes just as if we had got to the edge of the world and saw the sea come leaping down the gap. The descent, which is here sixty feet in a mile, aids the illusion, for the vision is circumscribed and the eye cannot reach far up the river. This is a scene which one is loath to leave, but we must on to the hot sulphur spring, where the guardian, an old Scotchman, lights the inflammable gas for our inspection, and seems literally to set the water on fire. Smell enough, and of a similar kind, comes from the Thames in hot weather; so if the guardians of that not very crystal stream carry their systems too far, they may see the Thames on fire some day yet. When our guide had shown us all, I asked him about the farming in the neighbourhood. He was from Aberdeenshire, and had been thirty years here. "Oh, man," said he, "they're meeserable farmers. It would break your heart to see how they just scart the grun'. It's no very guid ony way, but they dinna gie't a chance."

A few days may be spent with great satisfaction in wandering about Niagara, the grandeur and beauty of which seem to increase from day to day. Mr. Church has fallen in love with the Falls. He has made them his mistress, and morning noon and night he watches her in every varying mood and loves her in them all. At the dizzy edge of the Table-rock he will sit sketching all day with his easel before him, and, to the terror of the whole neighbourhood which lost sight of them for twenty minutes, he kept the little steam-boat, which plies below the Falls, all that time battling with full steam, to keep her nose in the spray in the very front of the Great Fall, whilst he studied the effect from this new position. If a passion for his subject can secure success to an accomplished artist, we may confidently hope that Mr. Church's second picture of Niagara will even exceed the beauty and fidelity of the first.

Returning by railway to Hamilton, I continued my journey through Canada West. The country to Dundas, and onwards for forty miles to Paris, is undulating, and seemed an easier and more fertile soil. Very little of it is wholly cleared, certainly more than half is still unbroken forest. But the trees are immensely tall, and show the rapid growth which only a fertile soil could produce. Though this district is quite within the limit of the profitable culture of Indian corn, a small proportion only of the land seems to be occupied by that crop. Its great value is everywhere admitted, but on this description of soil its cultivation demands too much labour. The last grain crop can hardly have been great, for in very few instances indeed are any ricks to be seen outside the barns, and they are not capacious enough to contain large crops. There has been, in fact, a very short crop in Western Canada. The young wheat is already green, as it is found necessary to sow very early in order to have the plant strongly rooted before frost sets in. On the best land wheat and clover are often taken in alternate succession, the clover being ploughed down in the hot summer weather, when the weeds are easily and cheaply destroyed, while the clover gives condition to the land for the wheat crop. All agricultural operations in this country are necessarily governed by the high price of labour. The principle which guides the American farmer is to take the most paying crop which can be grown at the least cost of labour. Being the owner of the soil, he has no landlord to consult as to its management, and he regulates his cropping by the cost of labour and the value of the produce most in demand. In every country men differ in their ideas on these points, and here as elsewhere some farmers find it most profitable to have their land chiefly under tillage, while others in the same locality think they realise a larger net balance from stock farming. The wheat crop on the best soils yields here from eighteen to thirty bushels an acre, and year old South Down sheep, where well managed, weigh 20 lbs. a quarter.

Up to the time of the money panic of last autumn the value of land in the best parts of Canada West had been for several years constantly rising. There was now a complete lull, and prices were on the decline. Great hopes had been built on the opening up of the country by the two nearly parallel lines of railway, which, at different points, traverse Canada West, and connect Michigan with Lake Ontario. These were realised more rapidly than could have been expected, aided by the high prices caused by a series of short crops in Europe, which stimulated production immensely for a time, and led to wild speculations in land. I have instances before me of farms in this part of Canada, only partially cleared, which then sold at 15*l.* an acre, and are now hardly worth the half of that money. A light sandy loam of good quality, only half cleared, is still valued at from 7*l.* to 8*l.* an acre. It is this comparatively high price of land, in addition to the cost of clearing off the timber, that forces the emigrant westwards to a country where better soil, with equal facilities of transport, can be bought for less than the mere cost of clearing this of its timber.

Some fifty miles farther bring us to London, a very rising and flourishing town, overlooking the broad wooded valley of the Thames. The soil here is a light loam, very favourable to the production of fruit and vegetables, and of easy cultivation. Onwards to Thamesville, Bothwell, and Chatham, the land seemed less valuable, and three-fourths of it still in natural forest. As we near the lake, and the St. Clair River, it becomes marshy. There seemed no great source here for much local traffic for a railway, and yet the Great Western cars were crowded, but chiefly with through passengers to Detroit and the west.

The climate of Canada West is more favourable to agriculture than that of Canada East, the winters being shorter and not so severe. But Lower Canada is more free from fever and ague. The diseases caused by malaria, (which is the most fruitful origin of disease in North America) become much less important or fail altogether in a poor cold country, for they are the product of rank vegetation on a rich soil, nourished by extreme heat and humidity. In proportion to the natural capacity of the country in North America to maintain a large population is its greater liability to such disorders. The rich alluvial soils on the river sides, whether in Canada or the United States, south of 45° of latitude, are equally subject to fever and ague, more or less intensified by the greater heat and humidity of the summer. The settler may make his choice with perfect certainty of the result. The poor soil, with niggard vegetation and harsh climate, is exempt from malarious fever, though favourable to colds and diseases of the respiratory organs. The rich and abundantly productive soil, where nature helps man in every way to make his labour profitable, nourishes also that exuberance of vegetation the decay of which produces malaria. But as these rich countries become populous the excess of vegetation disappears, and malaria diminishes. That has been the uniform experience of the Eastern States of America, as also of our own rich flats on the east of England. Certain it is that, notwithstanding malaria and its consequences population flocks towards the richer territory, and increases more rapidly in it. Canada West is richer than Canada East, and it is more populous; but there is a richer territory still farther west where labour is yet more productive, and, though in the present state of the country the risk of health is greater, it is ten times more populous, for men push on to the land in which they can most quickly and easily earn an independence.

Before leaving Canada and entering the United States at Detroit, it may be well to notice here a subject much agitated in the province at present, and also of national importance, viz., the opening of a direct route through the British territory between Canada and the Pacific. A company with most respectable colonial names is being organised to place a continuous boat and portage communication between Lake Superior and the eastern side of the Rocky Mountains. There are said to be two practicable routes, one exclusively through British territory, commencing at Pigeon River on Lake Superior, thence to Red River at Fort Garry, the British settlement there,—down the Red River to Lake Winnipeg,—250 miles through that lake to its northern extremity, and finally up the south branch of the Saskatchewan River, (which in volume and depth is said to be equal to the Mississippi above Dubuque,) and is navigable for a distance, measured as the crow flies, of 700 miles, to a point near the eastern base of the Rocky Mountains. This is said to be only eight days' journey from the gold districts of British Columbia. The other route begins at St. Paul's thence through the Prairie, 250 miles, to Graham's Fort on the Red River, from which point the stream is said to be navigable for steam-boats down to the British settlements at Fort Garry, where the two routes would join.

Hitherto the vast territory proposed to be opened up by this route has been represented to be unsuited by climate for settlement, and capable of producing only furs and hides. It has been so used by the Hudson's Bay and North West Companies ever since the French were expelled from Canada. And the urgent efforts of Lord Selkirk, so early as 1805, to colonise a tract which, from his personal knowledge, he estimated as capable of supporting thirty millions of people, were, by a combination of obstructions and disasters, finally extinguished in favour of the fur trade. More recent investigation has shown that the climate is not unfavourable to settlement, the summer temperature on the south branch of the Saskatchewan being the same as in the fertile region of northern Illinois and southern Wisconsin, while the buffalo winters in the belts of woodland on these northern rivers as safely as in the latitude of St. Paul's. It is said that all the grains of the cool temperate latitudes can be produced abundantly. It is the great resort of the buffalo herds, the presence of which in vast numbers sufficiently attests the plentiful supply of grass on the plains at all seasons. Those who take perhaps a sanguine view of the subject, assert that there is a country here four times the size of the British Islands, with a fertile soil, navigable rivers, and abundance of coal, now almost wholly unoccupied, which is perfectly adapted to settlement. They compare the state of this unnoticed territory to that of Europe at the period of the earliest Roman expansion, when Gaul, Scandinavia, and Britain were regarded as inhospitable regions fit only for barbarians, and anticipate an early rush of colonists from the old country to seize upon its natural advantages.

The evidence is sufficiently strong to show that this vast region ought no longer to be left unknown and unexamined. An exploring party is said to have already visited the sources of the south Saskatchewan, and there must be abundance of information regarding the whole territory in the possession of the Hudson's Bay Company. The descendants of the first colony, planted by Lord Selkirk in 1812, still maintain their ground on the Red River, but the vicissitudes they have experienced tend to modify the sanguine expectations, entertained by some, of the future progress of this country. Passing over their early years of contention with the North West Company, we find them in 1818 visited by a plague of grasshoppers, which nearly destroyed their crops. In the following year the destruction was complete. An eye-witness says that these insects were produced in masses, two, three, or four inches in depth. The water was infected with them. Along the river they were to be found in heaps like sea-weed, and might be shovelled with a spade. Every vegetable substance was either eaten up or stripped to the stalk. The colonists supported themselves during the winter by hunting; and in the following spring, by unheard-of exertion, succeeded in bringing seed from the valley of the Mississippi. From this time to 1826 they were unharmed, but in that year and 1852 their crops were submerged by river floods. The grasshoppers again returned in 1856, were most destructive to the crops in 1857, and threatened destruction in 1858, but were providentially stayed. Notwithstanding these repeated calamities, and though they have no market for their produce beyond a small demand by the Hudson's Bay Company, the Selkirk settlement is in a flourishing condition.

I had the pleasure of meeting at St. Paul's, Mr. Kitson, the mayor of that city, a man of great intelligence, who has been during the last fourteen years engaged in the fur trade at Pembina, on the American side of the boundary line at Red River. He has been at the British settlement during all seasons of the year, and knows no country in which the people live such an abundant and easy life. Their farms extend for thirty miles along the river. With very little labour they produce every thing that they require. The rivers and lakes swarm with fish, and the land has an abundance of game. But the summer is short and the winter long. The chief danger in the climate arises from early frost, which generally comes in September, but sometimes in August, and prevents the corn from ripening. The crops, however, are rarely lost from this cause. The grasshoppers are more to be feared. But the population, now about 4,000, increases. The people are chiefly Scotch and French Canadians. Mr. Kitson had never been on the Saskatchewan, but had often heard the valley described as very fertile. The soil in all the valleys he had ever seen in that country is a rich black loam, and near the rivers there is plenty of timber. Cultivation has been very little tried, as the Indians here live on game and meat alone. They raise no corn and eat none. Their daily food in all seasons is meat, the daily rations for a man being 7 lbs., and for a woman, 5 lbs. But, while Mr. Kitson believed that the valley of this river enjoys a milder winter climate than Montreal, he is doubtful whether Indian corn would ripen in it any more than it does on the Red River, which is considerably farther south, and where it is only grown as a garden plant. He thinks the route this way, and over the Rocky Mountains to Frazer's River is quite practicable. Indeed the country, as far as the Rocky Mountains, is said to be so level that the journey might be made the whole way in a carriage.

Nor would the fur trade of the Hudson's Bay Company be materially interfered with if their lease of this valley of the Saskatchewan and the Red River should not be renewed. They derive all their best furs from the Mackenzie River, and the vast territory to the north and east. Buffalo skins are their chief produce at present from this rich tract of country, and these are of minor importance.

But the English people have a duty to perform to themselves in this matter. If they are to hand over to Canada the absolute property in this great territory, Canada should be made to pay all claims for compensation to which the Hudson's Bay Company may be entitled. Why should we be taxed for an object in which they are chiefly interested? If the territory is not worth more than the sum to be paid as compensation, why take it from the Hudson's Bay Company at all? Speaking in the interest of England alone, it seems to me that it is better for us to have an English Company sharing among themselves in this country the surplus profits of that region, than that we should pay a large sum in compensation merely for the pleasure of transferring the property so acquired to Canada. The Canadians are the best judges of its value, and the most capable of developing it. If it turns out to be so rich as it has been represented, it will be their interest to encourage emigration to it. And it is equally their interest with ours to open a route to the Pacific, for the profits on the transit will be theirs, and the more the country becomes known and accessible, the more speedily will it add to the resources and wealth of the province.

It is indeed hardly conceivable that a country so far in the interior can be profitably settled, so long as there is abundance of better prairie land to be found 1,000 miles nearer home. And there is a difficulty which has not been adverted to, the hostility of the Indians, a very warlike race here, who have hitherto baffled all attempts of the Hudson's Bay Company to form agricultural settlements.

LETTER III.

Michigan.—Ferry Steamer.—Detroit.—State Agricultural Show.—Railway Hints for Home, to Prevent Dust, and communicate with Driver Shelter for Engine Driver.—Illinois.—Extent of the Rich Valley of the Upper Mississippi.—Chicago.—Its wonderful Progress.—Railway Investments.—Development too rapid.—Encouraged by high Prices of Agricultural Produce.—Money Panic succeeded by Failure of Crops, and unhealthy Season.—Immigration suspended.—Capacity of Country for rapid Improvement.—View of the State of Illinois on a Line of Seven Hundred Miles.—Settlers from Vermont.—Galena.—Dunleith.

At Windsor we crossed the American boundary line to Detroit, in the State of Michigan, by the St. Clair River, a deep crystal stream, nearly a mile broad, flowing with a gentle current of three miles an hour. This is the Bosphorus of North America, by which the navigation of the St. Lawrence and the Lower Lakes finds access to the three great inland seas of this continent. The ferry-steamer by which we crossed was large enough to accommodate 600 or 800 passengers with all their baggage, and, in the saloon on the upper-deck, tables were spread for supper, of which probably 100 partook. This forms a convenient resting and refreshment room for through passengers, who purpose continuing their westward journey by the trains ready to start from the other side.

Detroit is a very handsome town, finely situated on the river. It was laid out by a mathematical genius, who has succeeded in producing a very elegant, spacious, and conveniently arranged town. The old French farmers, whose original settlements stretched in long narrow strips back from the river, have all become extremely wealthy by the sale of their little estates, which are now converted into the most valuable town lots.

The Michigan State Agricultural Fair was about to be held. The show ground was enclosed and subdivided much in the same manner as at the agricultural shows in England. Very tastefully decorated booths were erected for the exhibition of flowers, and also of the fish which are caught or bred within the limits of this State, and which were exhibited in miniature ponds as at the great show at Paris, in 1856. The whole show ground was traversed by a wide carriage drive, which was to be used as a race-course for the trial of trotting horses, and for matches by lady equestrians, probably the most attractive feature of all Western State fairs. There were a few good short-horns in the yard, but the stock was chiefly of an inferior mixed breed. The best sheep are of the merino breed, and the wool produced in this State is reckoned good.

As I passed very rapidly through the State of Michigan, the line traversing a partially cleared country, the soil rather sandy, but picturesque, undulating and well watered by clear streams, I mention only a few points of practical interest. The first is in reference to railway comfort. Though the weather was extremely dry, and we travelled with open windows over 280 miles of dusty country at good speed, no passenger was annoyed with dust. A simple contrivance is adopted on this line which, in dusty weather, should be introduced into England. A thick canvass cover is stretched on a frame along the bottom of the whole train, covering the wheels and all the open space between the carriage and the rail, so that the dust as it rises is carried off and out at the end of the train in a constant stream. There is no practical difficulty in it, and the additional comfort to passengers is so great, that I can confidently commend the plan to railway directors in this country. Neither do the Americans seem to have any difficulty in carrying, by a line along the inside of the roof, a continuous communication from every carriage to the engine driver. This is in general use both in Canada and the States, and when any change took place in the arrangement of the carriages, the connection of the signal rope was effected without any difficulty or loss of time. There is another little matter which must have often struck railway travellers in England; the unnecessary exposure to weather of the engine driver and stoker. In the coldest, wettest, and stormiest nights these two men, upon whose care and consciousness the safety of the whole train depends, are whisked through the air at enormous speed, without any shelter or cover except an upright piece of iron with glass in it to protect their faces when looking out ahead. In America, these useful officials have a roof over them, glazed on the front and sides and open behind, within which they can carry on most of their duties, without unnecessary exposure, and from which they can keep a good look out, without being frost-bitten.

It was night when the train reached the first limit of the Great Prairie country, for a glimpse of which through the darkness, I strove anxiously but in vain, during the last hour of the journey to Chicago.

I had now reached the new capital of "that Western World," as Washington described it, which Penn prophesied would yet make a glorious country. The valley of the Mississippi above Cairo, comprising on its eastern bank Illinois and Wisconsin, and on the west, Missouri, Iowa, and Minnesota, embraces probably the greatest tract of fertile land on the surface of the globe. In total extent it exceeds England and France together, with the kingdom of the two Sicilies thrown into the bargain,—it is more than equal to Prussia and the whole Austrian empire,—even Spain and Turkey combined, would require the territory of the Ionian Islands to place them on a par with it. And this vast territory is not only intersected by numerous lines of railroad, which give it direct access to Montreal, New York, and Philadelphia, but on the north, by means of the lakes and the St. Lawrence, and on the south by the Mississippi River, it possesses a continuous water communication with the Atlantic.

Nothing can illustrate more forcibly the vast natural abundance and resources of this splendid country than the history of the grain-trade of Chicago. An Indian village in 1820, this place has become a great city with upwards of 120,000 people, with wharves and granaries for miles along the river canal which opens into Lake Michigan, and with streets, public buildings, churches, and private dwellings that may vie with those of London itself. The stores on the principal streets are equal in size and architectural elegance to the new row of fine buildings which leads from Cannon Street into St. Paul's Church Yard. There are numerous stands for hackney-coaches, and various lines of omnibuses ply along the streets. And Chicago is actually the centre of more miles of railway, completed and in operation, than London. Yet it is only twenty years since the first shipment of some forty bags of wheat was made from it. In 1837 its exports amounted to about 100 bushels of grain, in 1847 they had reached 2,243,000 bushels, and in 1857 upwards of 18,000,000 bushels. Chicago and all its wealth are in fact a property created by the profits arising in the mere transference from hand to hand of the surplus produce of but a small part of this wonderful country. Looking to Illinois alone, of which Chicago is the commercial capital and outlet, this surplus, great though it is, is capable of being increased tenfold, as only one-tenth of the fertile lands of this State are believed to be yet brought under cultivation.

But while no man who has seen the country can entertain

a doubt of its vast capability of further development, it does not surprise me that capitalists in London are disappointed with their railway investments here. In England we make railways to facilitate an existing traffic. Elaborate statistics are furnished to show the extent of the present business of the country proposed to be accommodated. But in the Western States of America railways are made for hundreds of miles through the wilderness, not to accommodate but to create traffic. You may often travel for miles through the open prairie without seeing a living creature, till the shrill whistle of the engine startles a solitary sand-hill crane or a covey of prairie fowl. An Englishman cannot at first imagine the possibility of a traffic to be found in such a country, adequate to the support of a railway. But the experienced American knows better the rapid rate at which population and produce increase in a rich open country, to which access is made. He points to the fact that six years ago there were only forty miles of railway in Illinois, the earnings of which fell short of 8000*l.* while last year the total earnings of the lines centring in Chicago exceeded 3,700,000*l.*

While, however, this is the fact, there cannot be a doubt that the development of railway accommodation has been too rapid, and has for the present outrun the immediate requirements of Illinois. This was encouraged by a state of circumstances which sanguine speculators did not perceive to be exceptional. A series of short crops in Europe, coupled with the cessation of supplies from the Black Sea during the Russian war, caused such a demand for the produce of Western America, as at once doubled the price of wheat, and thus rendered the cultivation of prairie lands enormously profitable. For even in the first year that such land is broken up, it can be successfully cropped with wheat, and to any extent for which labour can be procured. The temptation to settle on land, the very first crop of which in many instances realised more than double the cost of the land itself, was so great, that men with their families flocked from their poor farms in the Eastern States, and Lower Canada, in tens of thousands, to this land of promise. In two years the export of wheat rose from two to nine million bushels. The trade in timber, with which the new farms are housed and fenced, increased in like proportion. So sudden and extensive a demand on the carrying resources of the railways led many of them to provide working stock adequate for the traffic of a fully peopled and occupied country, their directors hastily concluding that this sudden prosperity would be continuous and progressive. And new lines were started in all directions by local land speculators, and by others who were not slow to profit by the flow of foreign capital which these golden prospects naturally directed to the West. The early anticipations of increasing traffic, upon the hope of which several of the great lines had raised their capital, were already more than realised; and distant shareholders saw no reason to doubt the sanguine anticipations of directors, who themselves might well have been deceived by such rapid prosperity.

In this inflated state the money panic of last year fell upon them. The price of wheat dropped a half, the farmers refused to sell, the rate of lake freights fell one half, and the receipts and traffic of the railways began to show a similar decline. The reduced prices continued during winter and spring, and were followed by a cause of even greater discouragement,—a season of extraordinary humidity, succeeded by sudden and excessive heat, the effect of which has in many places nearly destroyed the wheat crop, and in others reduced it to less than half of an average produce. So general was this unfavourable season in the north-west, that its effects are everywhere visible. After such a summer the autumn has naturally proved unhealthy, and bad crops and bodily ailments coming together, the spirits of the settlers have been sadly depressed. Bad news travel fast. Migration from the Eastern States is suspended, and foreign immigration has almost ceased.

It does not seem to me possible that there can be any real improvement in the traffic of the western railways before next harvest. The chief produce of the country which at present creates traffic, is the grain trade. There is but one crop in the year, and if that proves a partial failure, there is no help but to wait the result of another harvest. But nature is so bountiful in this country, and so small is the proportion of land yet under cultivation, that when the tide turns we may look for a rapid change. If with not more than a tenth of the good land of Illinois under a rude system of cultivation, the agricultural produce exceeded for a time the carrying capacity of the railways, what may it not become as the country becomes peopled and cultivated? With less than a million and a half of people, Illinois afforded in 1857 an amount of traffic which left a profit to the railways. A very few years, at her average rate of progression, will double that population, and at the same time double her agricultural produce. And if directors and shareholders will in the meantime act with prudence and patience their capital will soon again become remunerative.

Before examining particular localities in the State, I was anxious to obtain as it were a bird's-eye view of the country; such a general impression of its surface as would enable me to select points for special inspection. I therefore first traversed the entire state on the line of the Illinois Central Railway, from north-east to south, and from south to north-west, a total distance of about 700 miles. The State of Illinois extends from 37° to 42° 30′ north latitude, being thus nearly the same length as England, but further south, and on the same parallel with

Spain and Italy. This first journey occupied three days, the last day of September, and the first and second of October.

Immediately after leaving Chicago we enter on the prairie, which, near Lake Michigan, and for the first twenty miles, is low and wet, better suited for pasture and dairying than the cultivation of corn. The country then begins to rise, and in the next twenty miles the surface becomes dry and undulating; the soil a black mould, varying in depth from twelve to thirty inches, and resting on clay, or a mixture of clay and gravel. From this point to the Kankakee River, the first large stream we cross, the prairie is a series of long and gentle undulations, less abrupt than the chalk downs of England, but otherwise resembling them in general form and sweep. The character of the soil is very uniform, and the face of the wide open country is sparsely dotted with farm-houses. Where the prairie is unbroken, it is covered with long coarse waving grass, from three to four feet high; and in the hollows the grass is so high as to hide completely any cattle that may be grazing there. Before reaching Kankakee we pass through a settlement of 800 French Canadians, which has been growing for the last fifteen years. Each settler has about forty acres, and their farms are laid out along parallel roads at right angles to the railway. They exhibit signs of careful cultivation, and the village and church of the colony are prettily situated near the woods on the river side.

The town of Kankakee is finely situated on the river, fifty-six miles south of Chicago. Though there was not a house here five years ago, the population already numbers 3,500, with very good streets and shops, the centre of a rich agricultural district affording a sufficient traffic for a special daily train in and out from Chicago. The land behind it is a fertile, black, sandy loam, lying on limestone, excellent for oats and potatoes, and productive of rich grass.

Crossing the river, which is a broad clear stream, as wide as the Thames at Richmond, running between limestone cliffs clothed with timber, the road traverses a continuous prairie, more or less dotted with houses and farms for the next seventy miles. This is all a good range of country, and though the railroad frequently runs in a perfectly straight line for many miles, the surface while rather flat is very seldom a dead level, as may be at once observed by the varied depth of the cuttings and embankments all along the line. At every eight or ten miles we pass a station round each of which a town is rapidly springing up, very often with a steam flour-mill in its centre capable of manufacturing 150 barrels of flour a day.

At Urbana, 128 miles south of Chicago, there is a flourishing town and station, the population numbering near 4000. I saw a peach plantation in this neighbourhood which was said to be in some seasons extraordinarily productive and remunerative. High prices are paid by the graziers here for the best breeds of cattle to improve their stock, one man whom I met at the station having last year paid 500*l.* for a short-horn bull from England. The soil is very black and rich looking. Generally, even on the flattest prairie, groves of timber are visible somewhere on the horizon, but they become more frequent after we pass southward of Urbana, and until Mattoon is reached, a few miles from which, and at about 180 miles south of Chicago, the general level of the country falls about eighty feet. This forms the termination of the line of black loamy prairie, the grey wheat-soils of southern Illinois now commencing. The open prairie becomes narrower, and the woods, which are everywhere found along the beds of the rivers and streams, seem to be within little more than a mile apart from each other. The soil is more silicious than the black soil of the upper prairies, and better adapted for winter wheat, of which it seldom fails to produce good crops of fine quality. It is also considered good for grazing cattle; but is not so prolific of Indian corn or oats, nor so suitable for potatoes or sugar-beet, all of which grow very successfully on the black prairie. The face of the country, however, is more picturesque, and the woods more diversified, the white oak growing to a great height. There is also abundance of coal and building stone in this portion of the State, and the winter climate is occasionally so mild that in favourable seasons cattle can live the whole year on the prairies, with the aid of little or no fodder. From this point to Centralia, where the junction is made with the main line of the railroad, and onwards to the south as far as Desoto, which is 301 miles south of Chicago, the same whitish grey prairie soil continues. The country near Duquoin, a station on the line, is all underlaid with coal, in seams from five to nine feet thick, at a depth of seventy to eighty feet. It is easily wrought, but at present there is not much sale for it, as the country is very thinly settled, and there is no scarcity of wood. In the whole country, for nearly the last 150 miles, there was scarcely a settler four years ago, but so rapidly has settlement followed the opening of the railway, that it is estimated that half a million of acres of land have already been brought under cultivation along this part of the road.

From Desoto to the southern boundary of the State the country is all hills and hollows, rocky and wooded, with good farms interspersed. The climate is very mild in winter and hot in summer, and admits of the growth of all kinds of fruits and tobacco. It produces white wheat of the finest quality, and peaches and other fruits are sent in large quantities for the supply of the market at Chicago. This is one of the earliest soils in the Union for the ripening of wheat, the new crop from which may be sent to the northern and eastern markets before their own harvests are ready.

I now retraced my course by the same line to the junction at Centralia, but went northwards from that point by another line, nearly through the centre of the State, meeting with the same characteristics of soil as were noticed on the journey southwards. Near Tacusah there is another considerable settlement of French Canadians from Lower Canada. On again reaching the black prairie, after having been for some time accustomed to the whitish grey soil of the southern prairie, it seemed to me that the land looked richer and the grass greener. But we were now traversing the richest part of Illinois, and for 100 miles north of Tacusah the whole country is very fine, much of it settled and enclosed, and dotted with houses, as far as the eye can see. The cultivation is on a larger and more regular scale, the Indian corn and wheat both showing evidence of more careful management. Hay and corn ricks are more numerous; woodland is to be seen in all directions, and the country is altogether more undulating, rich, and picturesque, than any part of the prairie which I had yet seen. At Bloomington, which is a very rising town, with 7000 people, 10,000 bushels of grain are sent off daily by railroad to Chicago in a good season. The country here is chiefly settled by farmers from the middle States, Ohio and Kentucky. About thirty miles farther north, near the station of Minonk, a large colony of about 200 families from Vermont have settled. They sent before them a committee of their most skilful farmers to examine the Western States and choose the most suitable and advantageous position they could find. These men made a very careful inspection of Illinois, and other States farther west, during a four months' tour, and came to the conclusion that no other locality which they had seen presented so great a combination of advantages as this. They bought altogether about 20,000 acres, upon which they have been settled for the last three years.

At La Salle we cross the Illinois river, and have now reached the centre of the coal region of the northern part of the State, a busy populous district, in which the population has increased five fold during the last fifteen years. The value of land has increased in a much greater ratio, land near the station, which then sold at 10*s.* an acre, being now worth 10*l.*

At Mendota, about ten miles farther north, the country, which is all open prairie, is well "settled," and the people look unusually lively, healthy, and well fed. White clover may be seen growing very luxuriantly along the railway banks where the natural prairie grass has given way. The same kind of country continues for the next twenty-five miles to Dixon, which is a very handsome town of about 5000 people, finely placed on both sides of the Rock River, a broad navigable stream, flowing at the bottom of shelving wooded banks. For some miles north of Dixon the road runs up the river bank, skirting the woodland, and then emerges on a tract of open undulating prairie, where large farms with corn fields stretch out apparently for miles on either side. This continues for the next thirty or forty miles. In this northern part of the State the air is much cooler than in the south, and the winters are more severe. Cattle require six weeks longer of winter provender. Indian corn is not so productive by one-fourth as it is in the rich midland portion of the State, and winter wheat is so precarious that the spring-sown variety is chiefly cultivated. But this district is admirably suited for oats and potatoes, and for summer grazing. We have now reached Freeport, a flourishing town of 7000 people, on the Pecatonica river, northwards of which, for the next forty or fifty miles to near Galena, the prairie soil is thinner and more rolling, but covered with white clover wherever the natural grass has given way. This terminates the prairie land.

Galena is the great seat of the lead mines in America, and

yields annually about thirty million pounds weight. It is a large and thriving town, situated on the banks of Fever river,* which is navigable to the Mississippi, some few miles distant. The river smelt noxiously at night, and the principal trading streets lie along its bank. But the residences of the people are prettily scattered up the hillsides on both banks, and the inhabitants themselves, notwithstanding the ominous name of the river, think there are few places in the State to compare with the town of Galena. From Galena to Dunleith on the Mississippi, and near the north-west boundary of Illinois, the country has no interest of importance to a farmer. It is chiefly woodland, and, where open prairie, it is already "settled" and under cultivation.

* This is *Bean* river of the French. The name has no reference to any malaria arising from the stream.

LETTER IV.

General View of the State of Illinois.—Comparison of Soil and Extent with England.—Dunleith to Mendota.—Vast Wheat Fields.—Experience of a Scotch Carpenter.—Farming by Shares.—Cost of Farm-houses.—The River Illinois.—Coal Lands of La Salle.—Corn Starch Factory.—Bloomington.—Settlers from New York State.—Account of his Operations by Pioneer of Settlement.—Unusual Failure of Wheat Crop.—Discouragement caused by this.—Temptations of Credit System.—Instance of Purchase and Cost of making a Farm.—History of an early Settler.—The Banking System of the Country.—Profits of Banking.—"Shin Plaster" Banks.

I CANNOT hope in the preceding description to have conveyed to my reader more than I myself received in this hurried ride, namely, a general impression of the main features of the country, and an idea of an almost endless extent of fertile soil. Some time was required, and a careful study of the map, before even the outline features of this extensive country became lucidly fixed in my mind. I had first gone more than 300 miles due south of Chicago, and had then turned back, and, by a more westerly line, had run about 450 miles north, through the centre of Illinois to its north-western boundary at Dunleith on the Mississippi. To give a homely and at the same time pretty accurate idea of its extent, and bearing in mind that England and Illinois are nearly equal in size, let the reader imagine himself starting at Newcastle and proceeding by York, Newark, Peterborough, and Bedford to London, and then on to Brighton,—there let him turn back, retrace his course to London, and then take a north-westerly route by way of Rugby, Stafford, Manchester, Lancaster, Carlisle, and so on to Glasgow;—let him imagine the whole of this extensive country, with the exception of that portion between London and Brighton, to be an undulating plain, underlaid in various places with extensive deposits of coal and iron;—between London and Brighton let the country appear to him to be covered with timber, with a climate and soil peculiarly favourable to the cultivation of fruit and grapes, and to the production of the very finest quality of white wheat;—let the entire area from London northwards to Newcastle on the one side, and Glasgow on the other, represent "the prairies" of Illinois,—open steppe-like lands, covered with coarse natural grasses, with scattered copses of timber on the ridges and along the watercourses, and abounding in every element of fertility. He will thus be better able to realise the appearance of this vast open undulating plain, than which there is no other in the temperate zone so uninterruptedly extensive and fertile.

I spent most of the month of October in making a more minute and detailed examination of farming on the prairie, and will now ask the reader to accompany me in my ride, before troubling him with the figures and conclusions at which I finally arrived. The railway shall transport us from point to point on the route, and a very light waggon or carriage invariably drawn by a "span" (as they are called here) or pair of light active horses, of great spirit and endurance, shall convey us hither and thither over the country in our inspection of the prairie farms.

Starting from the north-western point at Dunleith, the first halt we make is at Mendota, about seventy or eighty miles from the northern boundary of the state. It is the point of junction with a railway running westward from Chicago to the Mississippi, opposite Burlington in Iowa. The road traverses a rich district of prairie, extremely favourable to cultivation. I travelled nearly 100 miles through this part of the country, and found the soil generally rich and deep, and the white clover and sown grasses very healthy and luxuriant. Some corn fields are of uncommon magnitude: one vast sweep of 2,200 acres was all in new-sown wheat, a sparkling sheet of verdure in the morning sun. The towns, most of which are not four years old, are growing rapidly. I met a carpenter from Lanarkshire, who had been settled in the country for twelve years. Though he had made money he could not keep it, but he blamed himself for this, as every steady man, he said, who had come to this part of the country from Scotland had thriven. Wages were at present lower than he had ever known them, a journeyman carpenter receiving only 4*s.* a day, with his board. He had seen many instances of men getting themselves into difficulties by buying more land than they had means to manage and pay for. But there is a plan of going "shares," in which a prudent Scotch farm labourer meets with great success. He has a farm given him to cultivate, fenced and broken up, and seeded;—he performs the rest of the labour and carries on the farm, and pays his rent by delivering at the nearest station the half of the crop. This is an arrangement by which a steady man is sure to succeed, and the owner of the land is also well paid.

The carpenter was at that time constructing a small farmhouse of timber by contract. The foundation was of mason work, with large underground cellar, the inside dimensions of the building being eighteen feet by twenty-four, divided into two rooms and a kitchen, with side posts twelve feet high, boarded, lathed, and plastered, and roofed with shingle, complete for 40*l.* Last year the same house would have cost 60*l.*; but both lumber and wages have fallen since the money panic about a third.

Following the route southwards for some twenty miles, we come to the Illinois River at La Salle. This river discharges itself into the Mississippi after a course of 500 miles, during which it drains nearly all the centre of Illinois, increasing its volume by the waters of many tributaries. It has been navigated by steamboats for many years, and furnishes a cheap water communication for the interchange of products anywhere along the line of the Mississippi, from St. Paul's to New Orleans. My object in stopping here was to inspect the coal-lands, of which La Salle is the great centre. Taking waggon at Ottawa, we crossed the river and drove some ten miles across the prairie, through a good country—all occupied—till we reached the bed of another stream, called the Big Vermilion, where the coal makes its appearance on the surface. Here, from the strata on the river side, we entered a shaft which penetrates a seam of coal nine feet thick, so situated that it drains itself. Near this, some thin seams of cannel coal have been found. All this part of the country is underlaid with coal, which may be mined with the greatest ease. The surface is fine, rolling, fertile prairie; and there is abundance of limestone everywhere. The La Salle coal-field, as at present worked, produces about 1000 tons of coals a week, and is capable of any required extension.

Returning to Ottawa, I visited a manufactory for making starch from Indian corn. It is situated advantageously in a good country for purchasing the corn, and with every advantage of abundant water-power, and canal and railway communication. Three kinds of products are manufactured, one of fine starch, one of ordinary starch, and one for making puddings. Two pounds weight of corn yield one of starch. The corn costs a farthing, and the starch sells at the factory for 3*d.* a pound, so that this business should leave a good profit. There is a similar work at Oswego, in the State of New York, which has proved very profitable to its owners. The article made there has found its way to England, is now extensively used under the name ot "Corn Food," and is by many considered more palatable and nutritious than either arrow-root or sago. This use of Indian corn is another of the many excellent purposes subserved by that most productive of all grain.

Crossing the Illinois River at La Salle by a viaduct 2000 feet in length, and 80 feet above the bed of the stream, we had a fine view of the limestone bluffs which here rise to a height of 200 feet. After a run southwards of some sixty miles, I stopped at Bloomington, which is within the district of what is usually reckoned the richest territory in the State. At this flourishing and pretty town I had the good fortune to obtain the guidance of a gentleman of much intelligence and local experience, who unites in his own person the various functions of banker, lawyer, judge, and colonel. We spent the day in riding over the country, and in looking at the farms and talking with the settlers. They were men chiefly from the State of New York, and were all complaining of the last wheat harvest as a nearly total failure. One of the pioneers of the settlement thus told me his story. He came here four years ago, and was so much pleased with the land and situation, that he advised his neighbours in the State of New York to follow his example. He purchased 2,500 acres of as fine prairie as can be desired. The first two years everything was successful. He grew more than thirty bushels of wheat an acre on the newly broken land, and sold it for *5s.* a bushel. He was thus tempted to lay out the money as fast as he made it, in enclosing and breaking more. The autumn before last he sowed 800 acres with wheat; 600 of it was killed by frost, the snow that winter being so light as not to cover it. He ploughed this up and sowed again with spring wheat, which succeeded admirably. Last autumn he laid down 600 acres with wheat, but was somewhat late and out of season in getting it sown. The spring proved unprecedentedly wet, the wheat was late in maturing, extreme hot weather set in, and his wheat, which till then looked well, was in one week rendered nearly worthless.

The long-continued rains in spring had given them no season for oats, and Indian corn had for the same cause been planted out of season, with the land in an unfavourable state, and the breadth very limited. This crop, which should be in the ground early in May, could not this year be planted till towards the end of June. Fortunately they had not had early frosts, so that there would be a fair yield. There had been no such unfavourable season for seventeen years in Illinois, and he knew that the farmers who had recently settled in the country, and who had had only last year's experience, were much disheartened. But personally he felt no apprehension, as he had the utmost confidence in the natural fertility of the soil, which he did not believe could be exhausted. He had seen similar land in this State from which twenty-four crops of Indian corn had been taken in succession, without manure, and the last was a splendid crop.

The next settler was a young man, a graduate of Yale College, who had purchased a section of land (640 acres) three years ago at nearly *3l.* an acre. He had built a house, enclosed his land, and broken up the half of it; but the wheat crop of last year, to which he trusted for future funds, had proved an entire failure. He spoke despondingly of his future prospects, as he had, like many others, been tempted by the facilities afforded by the credit system of purchase in these Western States, to buy a much greater extent of land than his available means were adequate for. He said that they all counted on their wheat crop to "bring them out;" but, that having failed them, the whole country was straitened. He wished to sell out at *4l. 10s.* an acre, that he might have the means of paying off his debts, and repurchasing a smaller farm in the same locality, where his obligations and risks would be less.

All the other settlers I met in this quarter had the same tale of a wretched wheat crop. One had had 120 acres of wheat, which he examined with several experienced farmers a few weeks before harvest, and they agreed in estimating the probable yield at eighteen to twenty bushels an acre. When it was ripe he began to harvest it, but after cutting seventy acres he discovered that there was nothing but shrivelled husk in the ear, so entirely worthless that he not only desisted from cutting the rest, but set fire to all that was already cut, as well as that which remained.

While the more faint-hearted were discouraged by the untoward season, there were many instances of an opposite kind. A very frequent cause of failure I found to arise from the incapacity of the settler to avail himself fully of his position. The credit system tempts him to buy a large extent of land, every unused acre of which becomes at once a dead weight upon him. If a man buys 600 acres and has not the means of cultivating more than 60, the 540 acres are a dead loss to him. He has to pay either the price, or the interest of the price, of this large unproductive and, to him, useless extent of land. The produce of the 60 acres is called upon to bear, not only its own burden, but that of the nine-tenths which are idle. The lean kine thus eat up the one fat one. In prosperous seasons so great a pull even as this can be withstood. But the first strain breaks it down.

An example of an opposite kind will show a more correct system. A person last spring bought 640 acres of land in this neighbourhood. He enclosed the whole of it, had it all ploughed by contract, and sowed it with wheat. Not an acre of his purchase was left idle. It was all sown in good order and in good time, and the chances were that the whole of it would succeed. As every part of the work was done by contract, and would be so completed, I am enabled to show the exact cost of the whole operation, and the probable return.

A ring fence round the whole, of substantial boards and posts, cost,		£240
Contract price paid for breaking the whole, and putting in the seed,		260
Paid for seed wheat,		160
Contract for harvesting, thrashing, and delivering,		500
		£1,160
Price of the land, cash, viz., $10, or £2 an acre,		1,280
		£2,440

Cr.

Probable crop:		
20 bushels an acre = 12,800 bushels, worth 75 cents, or 3*s.*,	£1,920	
Value of the land after being enclosed and broken, viz., $12½, or £2 10*s.* an acre,	1,600	3,520
Profit the first year, if the crop succeeds,		£1,080

These figures were given to me by a man of knowledge and experience; but the contract prices at present are lower than usual, and the cost of fencing and breaking is thus below the average cost of these operations. Neither is there any charge for buildings, though that would not affect the balance, as the property would be by that amount the more valuable. And it must be remembered that the wheat crop sometimes fails, while the above satisfactory result makes no allowance for failures.

I here learned the history of one of the early and most successful settlers in Illinois. He and his brother, then young men of twenty and twenty-one, came to this State thirty-four years ago, having left Ohio after an unsuccessful adventure in cattle trading. They were in debi when they began business in Illinois. They arrived in what was then a wilderness, and pitched their tent on the spot where one of them now resides.

They brought with them a pair of oxen, a mare, some few household utensils, a waggon, and two axes. They stopped near a "grove," and built themselves a shanty or log hut. In addition to the general stock, one of the brothers had a fur hat which, after his arrival, he "traded" for a breeding sow. They worked away, gathered live stock, there being no limit to the liberty of grazing, drove them great distances to market, and accumulated money. As the money gathered they bought up all the land they could get at the government sales, at 5*s*. an acre, continued their stock farming, and now send 100 fat cattle weekly to New York market, during the fall season after the cattle are fat. One of the brothers is believed to have now, in this prudent way, acquired land worth a quarter of a million sterling.

The soil in all this district is a rich black sandy loam, lying in fine gentle sweeps, admirably adapted for carrying on with ease and economy all the operations of husbandry, and, to use the phrase of the country, is very "handsome" prairie.

On my return to Bloomington I had an opportunity of learning something of the banking system of the country. The law permits any man, or company, who can purchase 10,000*l*. worth of State stock to issue bank notes. He deposits this stock with the treasurer of the State as security for the liquidation of the notes, and is then authorised to issue his own notes to the extent of 9000*l*., which are countersigned by the auditor of the State. These notes he lends out to his customers at the current rates of interest. The notes are payable in gold on demand, and if payment is refused, the holder protests the notes and carries them to the State auditor, who is in that case empowered to sell so much of the deposited stock as may be requisite to liquidate the protested notes. Thus if the State stock is convertible at or near par, the note-holder is always safe: and the banker makes an excellent business of it, as he receives his 6 per cent. dividend on the deposited State stock, besides the interest and commissions which he can realise in lending to his customers the notes which represent the same money. In a country like this where every farmer is the owner of his land, and where conveyances of real estate and mortgages are managed in the simplest and cheapest manner, there can be no business either safer or more profitable than that of a banker. The loans are made chiefly on the security of real estate, and the rate of profit in these new countries, where land is cheap and productive, admits of 10 per cent. as the common rate of interest on such security.

That such a system of paper currency must occasionally lead to embarrassment is self-evident. For with a general run on the banks there must be a suspension, as the State securities would in that case become as inconvertible as the bank notes.

But though the fair and legitimate profits of the bankers are thus very considerable, plans are occasionally adopted for increasing these profits, which are reckoned here, though a little "smart," still perfectly fair. If a banker is in good credit he finds that his notes will circulate readily although not payable in the State in which he carries on his business. He may have bought the stock of some other State, Alabama or Florida for instance, lodged his stock there, and obtained the countersignature of the State auditor to the authorised amount of notes which he dates within that State, and where alone they are demandable in gold. He does not issue them there, but brings them to his usual place of business, many hundreds of miles distant, and then lends them out among his customers. When the notes come back upon him he requires a commission, not that he disputes his liability or the soundness of the notes, but because he deems himself entitled thus to add to his profits, on the plea that if gold was wanted the holder of the note would have to incur a certain amount of charge in sending it to the distant place of issue!

But there is a still more questionable kind of banking adopted by some smart men in this western world, though I did not meet with any instances of it in Illinois. It is denominated the "Shin Plaster" or "Wild Cat" banking system. This is the description given to me by a man, who spoke from experience, of the way to get up such a bank. You go into a State where the stock is below par, say at 70 or 80. You buy 50,000 dollars of that stock, lodge it with the State auditor, and obtain his counter-signature to your bank notes. This paper money you take into the wilderness, knock up a shanty, write "Bank" over it, and date your paper money there. The more inaccessible the place is the better, as your paper is demandable in coin only at the place of issue. Having performed these necessary rites, you bring your notes to some centre of business: they receive currency at once from the State auditor's signature, and as you are a sharp business man you lend them readily on mortgage of real estate at 1½ to 2 per cent. per month. There is little fear of your notes coming back on you for payment, as the place of issue is undiscoverable. Every man into whose hands they come is interested in keeping them afloat. By degrees they are worn out, and thus with ordinary luck you secure your own deposit with the State, and its representative, which, in the hands of the public, has gradually disappeared! However improbable it may seem, I was assured that such practices are to this moment followed; but of course they are utterly discountenanced by all bankers of standing and respectability.

LETTER V.

Springfield.—Appearance of Country.—Cattle Show.—Stock Farming.—Experience of a successful Farmer.—His Mode of laying his Farm to Grass.—Novel Implements.—Merino Sheep Farming.—Account of it by the Owner of a large Flock.—System of managing Prairie Land recommended.—Sowing Grass Seeds on Snow.—Valuable Meadow.—Price of Merino Sheep.—Superiority of Prairie to Timbered Country.—The Governor of Illinois.—The Public Officers of State.—Manners of the People.—Decatur.—Lost on the Prairie.—The American Settler.—Mutual Help.—Fences.—Pana and its Neighbourhood.—Settlement of French Canadians.

FROM Bloomington I proceeded southwards to Springfield, the capital, and not far from the centre of the state of Illinois. This is a fine town, with good streets and shops, and the neighbourhood is diversified by timber. It is like all other places in this part of the country, surrounded by the wide prairie. The view from the top of the State house very much resembles that of the plain of Lombardy as seen from the Duomo of Milan, except that there is nowhere a boundary of mountains. But there is the same far stretching plain, with trees in lines and groups, the timber becoming denser along the banks of the streams, which have cut out for themselves hollow passages winding about on the panoramic landscape spread before the eye. The inhabitants of the town, like those in the country, are not this season exempt from ague.

I visited the country cattle fair or show which was then being held in a field close by the town. The best short-horn stock were exhibited by Mr. Brown, a celebrated cattle breeder of this State, whose acquaintance I had the good fortune to make in the show yard. He exhibited a short-horn cow, bred by himself, six years old, which had had five calves, a large fat handsome animal, which would have been a prize taker at any English show. He showed also a three-year-old short-horn bull from Lord Ducie's stock, imported last year. The large stock farmers of the West, who are the really monied men, are taking great pains to improve the quality of their cattle by the importation of the best English blood. It is an excellent policy, and they are already abundantly reaping the reward of their enterprise. For though at this autumnal season, the prairie grass looks coarse and innutritious, a stranger has only to examine the cattle which are fed upon it to convince himself of its feeding qualities. And, as this grass is everywhere to be had here for nothing, the grazier consults his own interest by incurring some expense in improving the present breeds of cattle, and thus obtaining earlier maturity, better quality, and quicker returns from his extensive grazings. Of the cattle common to the country there were several specimens exhibited, of enormous size. One red and white ox with wide upturned horns, four and a half years old, measured 2,700 lb. weight. He handled well, though very strong in the bone and limbs. Another of 2000 lb. gross weight was reckoned on the spot worth only 14*l.* at the current price of beef, viz. about 2*d.* a pound dead weight.

Mr. Brown has been many years in the country engaged in farming. He farms largely, and believes that more money may be made, and has been made, in this State by stock farming than corn growing. Nor is this remarkable, inasmuch as grazing land on the prairies hitherto could be had for nothing, costing neither rent nor taxes, while corn land must be bought, enclosed, and cultivated, and labour has hitherto been expensive. However, till very recently there was no outlet for corn. Railways are rapidly altering the former state of things, and Indian corn is no longer unsaleable at 6*d.* a bushel. He has found short-horn stock the most profitable, which is no doubt chiefly owing to the high prices he is enabled to realise in the sale of well bred stock for improving the breeds of the country. But he has not found them so successful on the natural prairie grass, of which on his own lands he has no longer any. Though the prairie grass may be extirpated in time by close feeding, he has found it the best practice to break it up, and, after a course of tillage, to sow the land out with blue grass and clover. The blue grass is a rich thick succulent grass of a bluish colour, which grows with great success on the limestone soils of Kentucky, and is found to succeed admirably on the prairies when laid down as pasture. It improves every year, and yields feed for six months, besides half feed during the winter, whereas the natural prairie grass is in its best state only for the first four months after spring. Mr. Brown has all his lands now laid down in "tame" grass, as the sown grasses are commonly termed here. He keeps no stock except his thorough-bred short-horns, and lets his surplus grass for grazing at one dollar a month for each animal, during the summer and autumn. He feeds his own stock during winter on the pastures, giving them corn and hay in time of snow. As he can buy Indian corn in his part of the State at an average of 8*d.* a bushel, he has no doubt that this is the kind of farming which best suits Illinois. He had tried sheep, and found them to do well, but having no taste for them he keeps exclusively to cattle.

There were various novel agricultural implements exhibited in the show yard. Ploughs mounted on an axle, with high wheels, the only advantage of which seemed to be that a seat was thus provided for the driver. There were seed planters of ingenious construction, a circular self-cleaning harrow, which always goes round about while being dragged forward,—little hand machines for washing clothes upon, which are said to economise labour 100 per cent,—and a chain-bucket pump, an extremely simple, cheap and efficient article.

I drove a few miles out of town to visit the farm of Mr. M'Connell, who was recommended to me by the Governor of the State as a man of great intelligence, integrity, and experience. I walked and drove over his farm, examined his stock, and received from him very clear and distinct information. He is a practical man who has been all his life engaged in farming, and has fought his way up to a very comfortable independence. He left "the old country" in 1811, farmed in a small way for thirty years in the state of New York, where he first settled, and moved thence to Illinois seventeen years ago. He had always preferred sheep-farming, and brought his small flock of merinos with him. They have been remarkably healthy, increase one-third every year, and his flock now numbers 25,000. His fleeces average four to five pounds each, and the wool sells for 1*s.* 8*d.* to 2*s.* a pound. He bought his farm at 1*l.* an acre, and could now sell it at 10*l.*, as it is in a good position near the capital of the State. But he is so firmly persuaded of the rapidly growing wealth of this fine State, that he has no doubt of his farm being worth 20*l.* an acre a few years hence. He considers the land for 100 miles round Springfield to be the best in the world.

Mr. M'Connell sends his flock to the open prairies in April, places about twelve hundred under the charge of one shepherd, who tends them and supplies them with salt. They need no other food for six months. He brings them to his enclosed ground in winter, and gives them hay when they need it, and a little corn. His flock has never suffered from any epidemic, but on the contrary have been extremely free from disease. His original flock grew one-fourth in weight and size after being brought from New York State to this better soil. He prefers the merino to the South Down for this climate and soil, and has found from trial that the merinos yield as much mutton and far better wool. He imports pure merino rams from Germany and Spain to improve his flock.

Mr. M'Connell finds that by feeding prairie grass close with sheep, it, in a few years, gives way to blue grass and white clover—which come naturally of themselves and without being sown. But the plan he recommends for laying this land down into good meadow and pasture, is to break up the soil some time between the middle of May and middle of July; (a few days earlier or later may be tolerated, but not more, as if prairie land is broken out of season the labour is worse than lost.) Sow wheat in end of August, or 1st of September: the following season, after wheat, take a crop of Indian corn, which must be kept clean; after the crop is removed, level the ground well, and in February sow one peck of Timothy to the acre,—if on the snow so much the better, as the dark seeds attract the sun's rays, and gradually melt a passage for themselves to the soil below, and the moment the snow disappears, they, being already imbedded in the damp soil, spring up at once, and take the start of all other vegetation. Late in March add two pounds of clover seed per acre, and a good hay crop will be certain.—I can testify to the success of this management, as I walked over a meadow of many acres on this gentleman's land, on which there was ricked a crop of at least two tons an acre of very excellent mixed clover and grass hay. The aftermath was rich close luxuriant clover, on which a flock of lambs were grazing, just such clover aftermath as we should find in this country on good land after the first crop of hay. I thought it had been the first crop, but learnt to my surprise that the meadow had been sown out twelve years ago, that it had little manure all that time, had borne a crop of hay every year, and been fed close afterwards with sheep, during winter and spring, till the prairie grass grew. I have never seen land in Britain that would bear a close clover aftermath at a period so distant from the time of being seeded, and cannot withhold my belief in the fertile qualities of a soil capable of doing so. Mr. M'Connell has no doubt that the prairie land would benefit by the occasional application of manure, but he never met with any other soil so constantly productive without it. He has known the first wheat crop pay the price of the land, with the cost of fencing it, and all labour, and leave a small balance over.

With regard to sheep-farming, his opinion is that corn and hay should first be provided by a few years' cultivation, before going largely into a flock. The prairie grass will furnish summer keep at little or no cost, but provision must be made for the winter. Good merinos can be bought for 8*s.* to 12*s.* 6*d.* a-head in flocks. There is probably no kind of farming on the prairies from which the returns would be so regular and certain.

Mr. M'Connell had tried a timber country before coming here, and was very energetic in expressing his opinion of the superior advantages to a settler on the prairie.

When in the capital I did myself the honour of visiting the Governor, who lives in a handsome house provided for him by the State, who also grant him the modest revenue of 500*l.* a year. He was a distinguished soldier in the Mexican war, and had long been one of the Senators of Congress. He has the highest hopes of the future of Illinois, and he, like other men of character and position to whom I have put the question, expressed the belief that fever and ague in this State are on the decline, though from special causes there had this year been an exeeptional prevalence of both.

I visited also the State House, where the two branches of the State Legislature hold their sittings, and in which are the bureaux of the various state officers. The Secretary of State very politely showed me over the building; the State Auditor supplied me with documents showing the valuation and taxation of the state; and the Treasurer, who locks up the money and disburses it exactly like the clerk in a bank, for which he is paid a salary of 400*l*. a year, explained to me the rate of taxation in the State, the desire they all had to pay off their debt, the present increased rate to which they submitted for that object, the probability of a future decrease in the expense, and the general frugality of the management. There is a total absence of form and ceremony about these gentlemen, who are high officers of state. The Secretary of State acts also as librarian. He and his clerk conduct the public correspondence and business. While I was there a man, about thirty, with his hat on and his hands in his pockets, came lounging in, and, after listening to our conversation for a while, asked if this was the Secretary, because he wanted to get some information about an old county road of which no record could be found in his county, but which he "reckoned" would be posted up at the capital in the books of the State. The Secretary immediately went off to "fix" him about the road. In the same way the Auditor was at everybody's call, and the Treasurer also. The officers of State are not above doing their own work here.

If there is not much official ceremony, there is a total absence of it in the manners of the bulk of the people. The nasty habit of chewing tobacco, and spitting, not only gives them a dirty look, but makes them disagreeable companions. They eat so fast, and are so silent, and run off so soon when they have finished their meals, that really eating in this country is more like the feeding of a parcel of brutes than men. The food is both various and plentiful, but it is generally badly cooked and served.

Violent thunderstorms are not infrequent. Every house on the prairie is fitted with a lightning conductor, but I did not hear that accidents from lightning were very common.

Again taking the railway, I proceeded to Decatur, a station about thirty miles east of Springfield, and drove for a whole day through the prairie country in that neighbourhood. After driving a few miles through the enclosed farms which surround the town, we reached the open unbroken prairie, and turning short off the track on which we had hitherto been driving, we stood across the great plain which stretched out before us. The horses struck without hesitation into the long coarse grass, through which they pushed on with very little inconvenience, although it was in many places higher than their heads. It was not thick, and parted easily before them; then sweeping under the bottom of our waggon it rose in a continuous wave behind us as we passed along. The surface of the ground was firm and smooth. We had fixed our eye on a grove of timber on the horizon as our guide, and drove on for about an hour in a straight line, as we believed, towards it. But stopping now and then to look at the soil and the vegetation, we found that the grove had disappeared. Without knowing it we must have got into a hollow, so we pressed on. But after two hours' steady driving we could see nothing but the long grass and the endless prairie, which seemed to rise slightly all round us. I advised the driver to fix his eye upon a cloud right ahead of us, the day being calm, and to drive straight for it. Proceeding thus, in about half an hour we again caught sight of the grove, still very distant, and the smart young American driver "owned up" that he had lost his way. We had got into a flat prairie about five miles square; one of the horses stepped a little quicker than the other, and we had been diligently driving in a circle for the last two hours. We soon struck upon a track which led us towards the rising ground and among some new settlements.

One man here had entered to an eighty acre lot last spring, had built his house, broken about ten acres and sowed it with wheat, and had his little crop of "sod" corn gathered and stacked out of harm's way, close to his dwelling. The first care of an American settler on the prairie is to provide for the first winter. If he starts in May he ploughs a few acres up, and very commonly plants the Indian corn on it by making a slit with his axe on the tough upturned sod, into which he drops the seed. Rude though this preparation appears, it is generally followed by a crop, sometimes a very good one. Having thus started his "sod" corn, he constructs his house, and spends the rest of the summer in "breaking" the prairie in preparation for a wheat crop, and in cutting and making some prairie hay for the winter provender of his live stock. He also plants a few culinary vegetables and potatoes. In the end of August he sows his wheat, and, when that is completed, he harvests his "sod" corn. This keeps him out of the market the very first winter, as it is often made to suffice for the food both of the family and the live stock. "Hog and Hominy" is not infrequently the only food that the settler has to set before his guest during the first year of his possession. And though homely it is wholesome. When the crop of Indian corn is secured, there is time to begin making fences. The neighbours have a mutual interest in this and assist each other. The fences are made of posts and sawn pine timber; the posts of cedar, seven feet long, cost *3d.* each, and both posts and rails are prepared in the forest, so that the settler buys them ready for his purpose, at either the nearest railway station or grove of timber, whichever happens to be most convenient. The holes for the posts are not dug out as with us, but are bored with an auger made for the purpose, and the work of fencing thus goes on with much neatness and regularity, and the fences, being all made in the same manner and with timber of the same dimensions, are very uniform and substantial. At this settlement we found the owner with four of his neighbours all busy in the work of fencing, one boring, one driving in the posts, and the others sorting and nailing on the rails.

The "snake" fence, which is common in all the timbered parts of America, is seldom met with on the prairie, and there only in the neighbourhood of a timber grove. It is a very substantial and excellent fence, but consumes too much timber in any country where that article is somewhat scarce.

In this day's ride, all the older settlers with whom we met, complained of the wheat crop as a failure this season, but the Indian corn was pretty good. One man who had settled here two years ago on good land, for which he then paid 30*s.* an acre, offered to sell it to us, with his "improvements" as they are called, viz. his house and a little bit of enclosure which he had made, at 62*s.* 6*d.* an acre. He was a considerable distance from a railway station.

My next stop was at Pana, about thirty miles farther south, where a junction is made with a line of railway which leads to the Mississippi, opposite St. Louis. From this point I traversed the country some fifty or sixty miles, and found the prairie in many districts almost unbroken. Here and there patches of unenclosed corn are seen, and sometimes incipient towns. The face of the country is generally beautifully undulated, with groves of timber in sight: the soil of blackish colour on a grey subsoil. It seemed a very desirable locality, and commands the market of St. Louis as well as that of Chicago. A French gentleman, a sugar planter in Louisiana, three years ago bought a large tract in this quarter, of about 25,000 acres, at 46*s.* an acre, which he is settling with a colony of French Canadians. He brought 400 people the first year, and nearly as many more the next. He sells to them in small lots at 66*s.* an acre, and it is said that the settlement is likely to succeed. The difference in price is not all profit, as he incurs sundry outlays in starting the settlers.

LETTER VI.

Pana to Centralia.—The Grey Prairie.—Best Wheat Soil.—Fruit.—Tobacco.—Vines.—Silk.—Rich Mineral District South of Centralia.—Lines of Communication with Ocean by New Orleans and Chicago.—Probable Market for Wheat in Cuba.—Description of Grey Prairie.—Value of Oxen.—German Settlement.—Large Purchase of Land by Kentucky Grazier.—His Plan and Prospects.—Farina.—Trading Spirit of the People.—Urbana.—Complaints of Wheat Failure.—Peach Growing.—Large Grazing Farm.—Management of Stock.—Uniformity of Soil.—Coldness of Weather.—Steam Plough.—Machines for economizing Manual Labour in greater Demand.—Bement.—Kentucky Settler.—His Plan of managing Eight Thousand Acres.—Onarga.—Its Neighbourhood.—Dairy Farming.—Artesian Wells.—Kankakee to Momence.—Price of Land.—Broom Corn.—Country from Momence to Monee.—Management and Produce.—Monee to Chicago.

From Pana I took the railway to Centralia, a station about sixty miles further south, and, in nearly a straight line, sixty miles east from St. Louis on the Mississippi. It is the point of junction of the main line of the Illinois Central railway with its branch to Chicago, and is about 100 miles north of the southern terminus of that line at Cairo. The surrounding country is the grey prairie soil of southern Illinois, which produces the finest quality of white wheat in the State, but is not so prolific of Indian corn or oats as the black prairie already described. It is, however, a superior fruit country, and possesses a climate suitable for the culture of tobacco, vines, and even of silk, though the last branch of industry has made no progress. Of tobacco there is produced annually nearly a million pounds weight, and the crop of fruit is valued at 200,000*l.*

But the whole country for the next thirty or forty miles is also underlaid with valuable minerals, which at no distant day will be highly prized. Limestone, marble, freestone, flag, slate, iron, and coal are all found here. The seams of coal vary from two to nine feet in thickness. The Du Quoin coal is of a glossy jet black, ignites rapidly, does not clinker, and yields a small amount of ash. In chemical composition it closely resembles the best steam coal of Ohio and Pennsylvania. It has many properties which place it in the front rank of western coals,—freedom from sulphur,—cleanliness when employed as a domestic fuel,—firmness to bear distant transportation,—and readiness in coking, yielding a large percentage of fixed carbon.

The distance of this part of the State from the great corn market of Chicago, nearly 300 miles, has hitherto retarded its settlement. But it is favourably placed for the cities of the Mississippi, and is nearer to the ocean than Chicago. When the line of railroad is completed throughout to New Orleans, which it is expected to be within a year from this time, southern Illinois will be brought within from 500 to 600 miles of sea-going ships at New Orleans, which is little more than half the distance by railway from Chicago to the Atlantic. And if the market of Cuba, where little or no wheat is grown, should be thrown open to America, a circumstance every year becoming more probable, the flour of southern Illinois will form the main supply of the Havannah market. These are considerations worthy of being kept in view in forming an estimate of the comparative value of different parts of the State.

From Centralia I drove through the country, first southwards for about ten miles, and afterwards to the north between twenty and thirty miles. The aspect of the country from the road is very different from the impression one is apt to receive in passing rapidly over it by railway. Instead of being very uniform and flat, as a stranger is apt to think it, there is much undulation; so much indeed that I am inclined to think that when the country shall be fully occupied, the various farms fenced, and numerous plantations made, it will lose the distinctive character of prairie, and assume the ordinary aspect of a rich well-clothed rural district. Nor is the prairie much more bare of wood even at present than many of the best arable districts of Scotland. Along the hollows scooped out by the rivers and streams there is always woodland. The woodpecker, prairie fowl, and quail are seen in abundance. The hickory tree yields nuts, the maple sugar, and hogs are turned into the woods to eat the mast. On the open ground the road, which is a mere track over the prairie, is constantly undergoing change, for the new settler puts up his fence on his boundary line, right across the track. The traveller must then strike a fresh track for himself. Orchards, chiefly of peaches, are everywhere being planted near the homesteads. One farmer, who had been nine years in the country, told me that he and his family cropped eighty acres, that the average yield of wheat was twenty to twenty-five bushels an acre, and of Indian corn forty. He would probably fatten forty hogs, worth 40*s.* each. He had a flock of inferior merino sheep in rather low condition. But the cattle on the prairie were large and in good condition, with a good skin. Three-year-old oxen, large and in what we should reckon fair condition for stall-feeding, are valued here at not more than 4*l.*

I must now ask the reader to turn back with me in a north-easterly direction, by the branch line of the railway, towards Chicago. On this course we shall have a distance of 250 miles still to traverse. For nearly eighty miles north of Centralia the prairie continues of the same grey soil which I have just described, more silicious than the black soil, and therefore better adapted for winter wheat, hardly so prolific of Indian corn, nor so suitable for oats and potatoes. It is more picturesque, more wooded, but also, near the river bottoms, more liable to fever and ague.

Near Effingham there is a thriving German settlement. I met a Kentucky proprietor who, two years ago, bought 5000 acres in this quarter for 10*s.* 6*d.* an acre. He has begun to improve it by breaking up the prairie for wheat, with which he sows the land down with blue grass. His desire is to bring the land early into good grazing ground, and the Kentucky blue grass seems to be an object of adoration to Kentucky men: and yet either it, or the soil of Kentucky, or the climate, must be inferior to our best limestone lands in England, for he admitted that two acres of his best blue grass land were needed to fatten a three-year-old short-horn ox. He hopes to make this land profitable by using the open prairie with all his cattle for three or four months, while at its best, and then having the cultivated grass on his enclosed ground to feed his cattle, when the wild grass becomes too coarse and rough in the autumn. As a Kentucky grazier he had found great advantage by the introduction of improved breeds of cattle from England, his experience being that a three-year-old short-horn ox, on the same land, attained as good size and better quality than the unimproved breed of the country at five years old.

Farina is a new station surrounded by this fine grey prairie. Though not many houses are in sight there are a good many settlers in the district of which this is about to become the centre. The stationmaster, an active and intelligent young man, had within the week opened a store, in which he had large supplies of all the requisites of a farmer's household. His sales already reached 8*l.* a day. He was prepared to deal in every imaginable way, and for every imaginable thing. He bartered broad cloth for wheat, candles for hazel nuts, ribbons for apples; in fact nothing was brought to him that he refused to take at a price, and to pay for in kind. There seems to be a market for everything in the West, the spirit of "trading" is so thoroughly ingrained in the people. At every station is to be seen a large wooden store with the words "cash for wheat" conspicuously printed up. The daily quotations at Chicago are known by telegraph at every station, and the price, less cost of transport, risk and profit, are arranged without difficulty. The wheat this year on the southern prairie is worth twice as much per bushel as that of the northern black land.

From Farina I took the railway to Urbana, nearly 100 miles farther north, the intervening country being parallel with and much the same kind of land as that already described at Pana and Decatur. The town of Urbana is situated in the midst of a fine rolling black prairie, with a solid mass of some 6000 acres of timber as a background. I spent two days in driving through the prairie in the neighbourhood, taking a circle of about twenty miles. The wheat farmers all complained of the nearly total loss they had this year sustained in their wheat crop, and some large landholders, not farmers by profession, were so much alarmed by the loss of the crop that they had discontinued sowing wheat. One man spoke of peach growing as a matter of profitable farming, and said that he had produced on his own land here at the rate of 5000 bushels an acre. This was done, however, on a very limited scale. There was a steam thrashing mill at work here, the owner of which assured me that he could thrash with it 1000 bushels of good wheat in a day, and that he had thrashed 150 bushels in an hour. But in that case the wheat had been cut with little straw, and the yield was very prolific.

Pushing on through the long prairie grass for some five or six miles farther, we came to the land of a large cattle farmer, a celebrated Illinois grazier. He is the owner of several thousand acres of land, and has been so successful as a feeder as on one occasion to have delivered 100 cattle at Chicago in one lot, the average weight of which was 2300 lbs. I rode over his farm, and through one enclosure of 2500 acres, which was partly in natural prairie, partly in sown grass, and partly in wheat stubble, and part where the wheat had never been cut, as it was considered worthless. We rode backwards and forwards over this extensive field for some time, under the guidance of the manager, looking in vain for a herd of 250 cattle, which we at length came suddenly upon, all lying among the long grass, and quite hidden by it until we were close upon them. They were all fine animals, rose up slowly, stretching and licking themselves, 100 of them being four-year-old oxen of great weight and fat enough for the butcher. But it was thought they would pay for farther feeding, and the intention of their owner was to feed the whole lot out on Indian corn during the winter. The cattle are fed on the ground where the corn has been cut, and they receive it in the straw, thus treading and apparently wasting it among their feet. It is not lost, however, as the rule in feeding is to put two hogs in to fatten with each ox, and the allowance required in fattening the ox, and the two hogs, is no less than 100 bushels of Indian corn. As the whole management is rude and rough one man is found capable of attending to 100 cattle and 200 pigs; but all he has to do is to open the shocks so that the cattle may get readily at the corn, and to supply them regularly with salt. The corn can be had in the field at this place, which is some distance from the railway, at about ten pence a bushel on the average, and at that price this kind of farming is found remunerative.

I continued my drive onwards through the prairie, the most of which was still unbroken and unoccupied. There seemed no difference in the quality of the soil, which possesses a remarkable uniformity, the only apparent distinction being in the greater flatness of some sections, and consequently the greater liability to injury from wet seasons, where the land has not a sufficient undulation to keep it free from surface water.

This day (middle of October) was bitterly cold, colder than I ever felt the wind at home, and we were glad to get off the bare prairie, and into the shelter of the woodlands.

We passed a steam plough which was moving itself along a prairie road to a farm where it was about to be tried. It was a rude-like implement, with six common ploughs fixed to a framework, which could be let down or raised at the back of the machine. Nobody could tell us whether it had succeeded or not, though certainly no land in the world is better suited to steam culture. But its general introduction here may be retarded by the low value of the food of working cattle. Working oxen can be kept on the prairie for absolutely nothing, and in winter may be fed on prairie hay, which costs very little labour indeed. And corn for horses is also very cheap. There is thus not the same necessity for saving ox or horse labour as in England. Machines which economise human labour are in far greater demand.

I saw also a rude kind of mole plough. It was simply a clog of wood fixed to a strong rope, which is drawn by a powerful team of oxen through the hollow parts of a farm, two feet below the surface, and which thus leaves a passage for draining off the water.

We stayed all night at Bement, a village and station on the Great Western Railway. Many of the people had been suffering from ague, and this, with the bad wheat crop and fall in prices, produced a considerable depression of spirits. A few miles west we came upon the farm of a Kentucky gentleman, who, with his brother, had bought 8000 acres of land, fine gently rolling prairie, which he was bringing under cultivation. They had 800 acres sown with wheat, but the crop of the previous harvest had been a failure, having yielded little more than six bushels an acre. The Indian corn was a tolerable crop. He is now sowing out his land as fast as he can, his present plan being to have it all in blue grass, timothy, and clover, except about 1000 acres, which is to be kept under Indian corn for fattening his cattle. He then hopes to be able to sell 1000 fat cattle annually, and, if he can succeed in this, his purchase will prove extremely remunerative. The blue grass I find everywhere spoken of as best adapted for the prairie. It is sometimes sown on the fresh prairie after the grass has been burnt off, the ground being first well harrowed. But though this occasionally succeeds, the process of breaking up the prairie and sowing it out after one or two corn crops is preferred. Turnips are grown here as an experiment with fair success. Sown after wheat harvest, the roots are now 2 to 3 lbs. weight. Large tracts of land are for sale here at prices from 40*s*. to 50*s*. an acre.

Returning from Bement to Urbana I drove through a fine rich rolling prairie country, the larger proportion of which is still open and unoccupied. A nurseryman has established near the railway line a very thriving and extensive nursery of various fruit and forest trees, the thriving condition of which sufficiently proves the capability of the prairie soil for the growth of fruit trees and ornamental timber.

Taking the railway at Urbana I again proceeded about forty miles farther north to the station of Onarga, a rising town on a fine prairie, which seems all dotted over with neat two-story houses. I visited a good many recent settlers in this neighbourhood, most of them men with no previous experience of a country life, and without any knowledge whatever of the practical details of farming. These persons were all disheartened by the failure of the wheat crop. But others again, who had been brought up to farming and understood their business, were hopeful and making every exertion to ensure success.

The first man we called on was a dairy farmer from the Eastern States, an intelligent practical man who thoroughly understood his business. He has had the same ill luck in his crop as other people, but knows that risk of seasons is one of those risks which farmers in all countries must more or less reckon upon. He has a dairy stock of thirty-eight cows, and makes the milk into cheese. He can sell his cheese on the spot at 42*s.* a cwt., which is not far short of the average price realised by dairy farmers in Scotland, where the *rent is higher than the price of land* in Illinois. He finds the natural prairie grass very productive of milk till the month of September. His cows yield him 2 lbs. of cheese each, daily, during the period of good grass; and they can be foddered very cheaply during the winter on prairie hay. He expects to improve his stock and returns materially, as he goes on, by providing succulent food for the autumn and spring.

The prairie in this district frequently rises to rounded hills, which though more picturesque than the long gentle sweep of what is termed a "handsome" prairie, is not so fertile. The soil is much more sandy. There are several artesian wells here. I saw one, the water of which was rushing up full, through a four-inch pipe from a boring 127 feet deep. It was iron tasted, but very wholesome, and is constantly pouring out at this rate.

The last station at which I stopped to examine the country was thirty miles farther north, at Kankakee, which is fifty-six miles south of Chicago. From this thriving town I drove for about twelve miles up the north bank of the river to a town called Momence, and thence struck right through the prairie for upwards of twenty miles to Monee, a station within thirty-four miles of Chicago. The first part of the ride as far as Momence was through a very fine dry rolling prairie, which comes down to the bank of the Kankakee river, a broad clear stream, running over a limestone bed, the banks of which are wooded and picturesque. Improved land sells here at 5*l.* to 8*l.* an acre. The Indian corn was good and well managed, and I observed several fields of broom corn, a tall plant, exactly resembling Indian corn and cultivated in the same way, but bearing its seed, which is like millet, at the top. It is largely cultivated for the manufacture of brooms, for which the seed-bearing fibres, which are tough, elastic, and flexible, are used. An acre of this plant is much more valuable than Indian corn. Manufactories are established in the State for making it up, and a crop which is in all respects suitable for the purpose yields sometimes as much as 20*l.* an acre.

From Momence to Monee we passed through the same description of dry black rolling prairie. The country is higher here, and the winters more severe. A settler told me that they had generally to fodder their stock for seven months, for though the snow did not lie long, the frost bound the soil eighteen inches down. In order to secure the safety of the winter wheat it is sown among the growing Indian corn, plenty of the stalk of which is left when the corn is reaped to catch and hold the snow, which thus shelters the young wheat from the intense frost. The average crop of wheat is twenty-five bushels an acre, forty to fifty bushels of Indian corn, and forty of oats. But this year oats proved a total failure. White clover and blue grass everywhere make their appearance among the prairie-grass, where that is closely pastured. Cattle thrive well on these prairies, and the natural hay cut in August or September, on the upland prairie, makes excellent fodder for both horses and cattle. But horses do not thrive so well in the summer on the prairie, they are so tormented by the horse-fly, which seems unable to make any impression on the thicker hides of the cattle.

Towards evening we reached Monee, and an hour or two

afterwards I completed my tour of the State at Chicago. About a fortnight later, on my return from the Upper Mississippi, I had an opportunity of verifying my first impression by traversing the country both in the former and in a new direction.

LETTER VII.

Soil and Climate of Illinois.—Nature of Prairie Soil.—Its Chemical Composition. —Rich in Nitrogen.—Wheat Culture and Produce.—Indian Corn.—Facility of Culture.—Oats.—Barley.—Sorghum.—Substitute for Sugar-cane.—Potatoes. —Stock Farming.—Prairie Grass.—Blue Grass.—Timothy.

HAVING now obtained the necessary information for forming an opinion of Illinois, I propose here to consider its advantages as a place of settlement.

SOIL AND CLIMATE.

The characteristic soil of this State is that of the prairies, of which it chiefly consists, and to which alone my attention was directed. They comprise many million acres of land, more or less undulating,—in their natural state covered with grass which is green and succulent in May, June, and July, and shoots up in autumn from three to six feet in height.

How the prairie formation originated it is unnecessary here to inquire. It is sufficient to know that we have a soil evidently of great natural fertility, which for thousands of years has been bearing annual crops of grass, the ashes or decayed stems of which have been all that time adding to the fertility of the soil. So long back as we have any knowledge of the country, it had been the custom of the Indians to set fire to the prairie grass in autumn, after frost set in, the fire spreading with wonderful rapidity, covering vast districts of country, and filling the atmosphere for weeks with smoke.

In the course of ages a soil somewhat resembling an ash-heap must have been thus gradually created, and it is no wonder that it should be declared to be inexhaustible in fertility. In Europe such tracts of fertile country as the plain of Lombardy are known to have yielded crops for more than 2000 years without intermission, and yet no one says that the soil is exhausted. Here we have a tract naturally as rich, and with the addition of its own crops rotting upon its surface, and adding to its stores of fertility, all that time. It need occasion no surprise therefore to be told of twenty or thirty crops of Indian corn being taken in succession from the same land, without manure, every crop, good or better, according to the nature of the season.

Externally the prairie soil appears to be a rich black mould with sufficient sand to render it friable, the surface varying in depth from twelve inches to several feet, lying on a rich but not stiff yellow subsoil, below which there is generally blue clay. This drift surface lies on rocks consisting of shales, sandstones, and limestones, belonging to the coal measures.

Its chemical composition has been ascertained for me by Professor Voelcker, consulting chemist to the Royal Agricultural Society of England, to whom I sent four samples of prairie soil for analysis, brought by me from different and distant points of the lands belonging to the Illinois Central Railway Company. The letter of Professor Voelcker, and a copy of the complete analysis, will be found in an Appendix. They bear out completely the high character for fertility which practice and experience had already proved these soils to possess. The most noticeable feature in the analysis, as it appears to me, is the very large quantity of nitrogen which each of the soils contains, nearly twice as much as the most fertile soils of Britain. In each case, taking the soil at an average depth of ten inches, an acre of these prairies will contain upwards of three tons of nitrogen, and as a heavy crop of wheat with its straw contains about fifty-two pounds of nitrogen, there is thus a natural store of ammonia in this soil sufficient for more than a hundred wheat crops. In Dr. Voelcker's words, "it is this large amount of nitrogen, and the beautiful state of division, that impart a peculiar character to the soils, and distinguish them so favourably. They are soils upon which I imagine flax could be grown in perfection, supposing the climate to be otherwise favourable. *I have never before analysed soils which contained so much nitrogen, nor do I find any record of soils richer in nitrogen than these.*"

Crops.

Though these soils are so rich in nitrogen, they seem to be too loose for wheat, which is undoubtedly a precarious crop upon them. The open prairie country is so wind-swept in winter that snow seldom lies long to any depth, and the young wheat is thus left unprotected to the frost. Should it escape that, it is liable to be thrown out by the rapid changes of weather in spring,—and if it is fortunate enough to escape both, it is sometimes destroyed, as it was last year, by its enormously rapid growth in forcing summer weather, growing as it does almost on a muck-heap. In such a season as the last, the prairie wheat crops of Illinois were injured precisely in the same manner as our own in this country sometimes suffer from a too heavy dose of guano, in a warm moist summer. The growth is too rapid, the vesicles of the stem burst, and the ear does not fill. I cannot doubt that Professor Voelcker indicates the proper remedy for this in the application of lime, of which these soils are comparatively deficient. It would consolidate the soil, render the wheat less liable to be hoven, and help to strengthen the straw, and render the growth less rank. There is abundance of lime in the country, so that the remedy is at hand, and will undoubtedly be applied under a more scientific system of agriculture.

Autumn wheat is the most valuable corn, but it is also the most difficult to be grown, for it has to withstand the unprotected severity of winter. The earlier it is sown after the 1st of September the more likely it is to succeed, and it is generally successful when sown on the first and second crops of a newly-ploughed prairie which has been broken in proper season. If any of it should be destroyed by frost, the ground is sown in spring with spring wheat, and this seldom fails. The crop varies from fifteen to forty bushels an acre, twenty being reckoned a fair average.

Indian corn usually forms the third crop, and if the land is kept clean by diligent horse-hoeing, this crop may be repeated as long as the farmer likes. It is undoubtedly the main crop of the prairie farmer: it never fails. In some seasons it is more productive than others, but the most ordinary care will secure a crop. Under good management the yield often exceeds 100 bushels an acre, and, in the middle districts of the State, fifty would be reckoned a moderate average. One great advantage of this crop is that a large breadth of it can be cultivated by one man and a couple of boys. Two men and a boy with four horses can till 100 acres. Instances have been known of 100 acres of Indian corn and 50 acres of wheat being all managed by a man with his two sons and their horses. There is no hurry in harvesting it. It can be cut at any time after it is ripe, and takes no injury by standing either uncut, or in the shock, for many weeks. When consumed on the ground by cattle the shocks are merely opened, and the cattle shell the corn for themselves. It is always convertible into money, either as corn or pork, and it is extensively used in distillation and in the manufacture of starch.

Oats are not so certain a crop, but they are extensively grown in northern Illinois, and average about forty bushels an acre. They are light compared with good Scotch oats, and more resemble the oats of Northern Germany.

Barley is a valuable crop when it succeeds, as it is largely used in the making of beer, for which a growing demand is springing up throughout America. I have little doubt that the application of lime to the prairie would render it a better barley soil. The crop averages forty bushels an acre, on suitable soils.

Sorghum saccharatum, or Chinese sugar-cane, is cultivated in every part of the State, as yet experimentally, for the production of sugar. The leaves are found very succulent and nutritious as fodder when taken off before the plant ripens. It grows precisely like Indian corn, and can be produced successfully on the best corn soils. It must not be sown near broom corn, as the plants hybridise and both are deteriorated. Some carefully conducted experiments show that the yield of sugar, per acre, from this plant has amounted to 1221 pounds, with seventy-four gallons of molasses, which is about two-thirds of an average cane sugar crop in Louisiana. It has been satisfactorily proved that this plant may be made a substitute for the production of sugar by white labour, should any circumstance hereafter curtail the cane produce of the slave States. The prairie soil is also admirably adapted for the growth of sugar beet.

Potatoes are a productive and valuable crop in the northern part of the State. They yield from three to seven tons an acre, which sell at from 30*s.* to 3*l.* a ton.

The rates of price for agricultural produce in America, as has been already shown, depend on the increase of the population, and the capability of the land to produce certain crops. Price is also very much modified by improved means of railway transit. Before railways were introduced into Illinois the expense of transport was so great, that no farmer, unless he lived near a large town, could cultivate any kind of corn profitably. Stock farming was then the most remunerative kind of husbandry, and the men who have become wealthiest in the State have made their money by stock farming. We shall now therefore devote a few sentences to the value of the different kinds of grass and provender produced in Illinois.

The prairie grass shoots up fresh in the month of May, and continues green and succulent till August. All kinds of stock thrive on it during this period. Cattle, which have been carefully wintered, and are turned out upon it in good condition, will become quite fat. Milch cows yield well upon it, and Merino sheep also thrive. After August it shoots up, and becomes comparatively hard and wiry. The most forward stock should then be placed on "tame" grass, but growing cattle of all kinds may be kept on the prairie till November. In August and September it is usual to cut as much prairie hay as may be requisite for winter provender. This is got very cheaply. I was told by a large stock-master that three teams a day, one cutting with a machine, one raking, and one stacking, might in these two months save as much hay as would winter 1000 head of cattle. I do not think that the natural prairie grass is nearly thick enough on the ground to maintain so much stock on a given extent of land, as our good pasture land in England. But at present that is not a question of much importance, inasmuch as unenclosed prairie can be had in most parts of the country for nothing, and when the population becomes dense enough to occupy all the prairie lands, these natural grasses will have disappeared and have given place to corn and cultivated grass.

The grass most generally preferred is the blue grass (*Poa pratensis*), which is indigenous on the limestone lands of

America, and will usurp the place of all other grasses on such soils in the course of years. It is said to yield a greater return of beef, milk, mutton, wool, or pork, than any of the cultivated grasses.

Timothy grass is not adapted for pasture, but is generally used for hay. It yields from one and a half to two and a half tons an acre.

LETTER VIII.

Average Prices of Agricultural Produce in Illinois.—Cost of Labour.—Cost of Indian Corn in England.—Cost at which Pork may be raised by it.—Profit of Farming in Illinois.—Detailed Example.—Lands of Illinois Central Railway. —Advantages of their Position.—The Company's Terms of Sale for Cash or Credit.—Exemption from State Taxes till paid for.—Comparison between Farming in England and owning Land in Illinois.—Capital necessary to start one Farmer in England sufficient for four Land Owners in Illinois.—Profits of Sheep Farming.—Lands farther West only apparently cheaper.—Great Opportunity for Farm Labourers of Character and Skill.—Farming by Shares.— Facility for investing Money in Land.—Even the Labourer can so invest his Savings from Time to Time.—Prospects of Emigrants from Towns.

WHAT are the profits of farming on the prairie lands of Illinois? That is the question of interest to the agricultural readers of this little book.

The average prices of wheat and Indian corn in Chicago, since 1850, and those of beef, pork, cheese, and butter, since 1854, have been :—

	s.	*d.*
Wheat, per bushel,	3	9
Indian corn,	1	8
Beef, per lb.	0	2¼
Pork,	0	2
Cheese,	0	4
Butter,	0	8

These prices may be reckoned, on an average, as about one half the value of the same articles in England. If the cost of production in the two countries were nearly the same, the value of land of equal quality should be twice as great in England as in Illinois. It is, however, thirty times as great, and in this disparity consists the advantage which a settler may hope to reap by buying land in Illinois.

The cost of production is an important element of price. Manual labour is 100 per cent. dearer in Illinois than in England; but the cost of keeping horses is 100 per cent. cheaper, and as a larger proportion of the work of the farm is done in America by horse power and machinery than in England, the cheapness of horse labour will fully compensate the prairie farmer for the dearness of manual labour. The cost of production, in so far as labour is concerned, is thus much alike in the the two countries.

I have already said that wheat has proved, during the last two years, a very precarious crop. It can usually be grown with safety as the first, and sometimes the second crop, on newly broken prairie. And it is also a pretty safe crop to follow grass land when first broken up after having been some years laid to pasture. But Indian corn is the crop of the prairie farmer, and there is always a market for it either by selling or consuming it in the fattening of hogs. If grown within 150 miles of Chicago, it may be carried by rail to that port, there shipped to Montreal, and thence to England, where it may be delivered at 25*s.* a quarter, after paying all expenses, and it will then leave 1*s.* 8*d.* a bushel to the grower. At such a price there would be a demand for any conceivable quantity, as it would be the cheapest food for horses, cattle, and hogs that we have ever had in this country. It was calculated by Mr. Lawes, as the result of experiment, that about 4 lb. of meal produced 1 lb. of pork. Indian corn meal at this price would cost less than ¾d. a lb., and the English feeder could thus produce pork at a cost of 3*d.* a lb. This would be a great boon to the English farmer, and leave a paying price to the producer in Illinois.

I have now before me four detailed accounts of farms of eighty acres each, all of which show a profit, besides paying for the land itself, from the first crop. But these cases were instances during the period of high prices in 1855–6. And the same may be said of all the detailed accounts which have been recently laid before the public. I propose, therefore, to offer an estimate based on the probable future range of prices, and, to facilitate calculations, will take 100 acres of land; the first crop wheat, and the following crop Indian corn. The wheat crop shall be cultivated by contract, the land fenced, broken up, sown with wheat, reaped and thrashed, and a labourer's house built, during the first eighteen months. The second and following crops shall be managed by two resident ploughmen, whose wages, and the cost of keeping their four horses, will be the only outlay.

Cash price of 100 acres of land, ...		£200
Contract price of fencing, breaking, sowing with wheat, reaping and thrashing, and building a labourer's cottage, and stable and shed, ...		250
Capital invested in the purchase of four horses, implements, and harness, ...		110
		£560
Second year, wages of two men, horse keep, taxes, and accounts,		200
		£760
Cr.		
First crop, wheat, 2,000 bushels, at 3*s.* 6*d.*, . .	£350	
Second crop, Indian corn, 5,000 bushels at 1*s.* 8*d.*, .	416	766
Surplus after the second crop, besides the value of the land and stock, ...		£6

The third year begins by the prairie farmer finding himself the unencumbered OWNER of his land, all fenced and improved, with a stock of horses and implements, and the whole of his original capital in his pocket. He may continue to crop his farm with Indian corn, from which he will reap very large returns on his capital.

The foregoing example has reference to a capitalist purchaser, not a working farmer. The 100 acres may be multiplied by any number for which there is adequate capital, and the results ought to be the same in proportion. There appears to be thus a very ample surplus in the way of annual return, whilst the value of land itself will probably treble within ten years from the mere growth of population.

But a working farmer will not only receive the same annual dividend from his capital, but will also take to himself the full rate of wages which is allowed for hired labour in this estimate. And he may, moreover, avail himself of the credit given by the Illinois Central Railway Company to the purchasers of their lands.

That Company have still 1,300,000 acres of land to sell. It is situated along their line of railway, chiefly within five miles on either side, and affords every variety of soil, climate, and situation to be found in the State of Illinois. They offer their lands at prices which, considering situation, quality, and terms of payment, are the cheapest I met with in America. Every facility for the transport of produce to market is at the command of a settler on their lands. At every nine or ten miles there is a station, with an electric telegraph, where the latest news of the markets may be learned; while there is usually a store at the station for the sale of produce, and the purchase of necessaries. Their terms of payment for the land are either cash with a proportional discount in the price, or a long credit with a low rate of interest. So confident do they feel in the increasing value of their land, that they readily leave the entire price of it as a mortgage to be repaid by annual instalments out of the produce of the land itself. Purchasers from them obtain the further important advantage of exemption from all State taxes until the whole instalments of the price have been paid off, and this usually extends over the first seven years.

Let us consider the advantage of this credit plan to the father of several sons in this country, to whom he may be anxious to give the means of starting in life. If he desires to place one in a farm in England, of 300 acres, he must provide him with a capital of 2,000*l.* But if, instead of making his son the tenant of another man, he determines to purchase a farm of the same extent for him on the prairie, he may pay the advance interest of the purchase money of the land, fence it, build on it, stock it, and sow the first crop for about 500*l.* Two years elapse before the first instalment of the price is due, and by that time, with good management, the land should have yielded enough to pay it, besides all the expenses of management. An intelligent, prudent man, with 500*l.* in his pocket, may rely on finding that sum sufficient to start him successfully on 320 acres of prairie land, if he avails himself of this credit system.

His position will be this. He enters into a contract with the Company for the purchase of 320 acres of their land, at the price of 50*s.* an acre. He pays 6 per cent. advance interest upon this, but he pays nothing further for two years. His first instalment, one fifth of the price, then becomes payable, and each year thereafter, till all is paid, another fifth. His account will stand thus:—

Two years' advance interest on price of land, ...	£48
Contract price of fencing 100 acres, breaking it, sowing with wheat, reaping and thrashing, and for building a house, stable, and shed, ...	300
Price of horses, implements, and harness, ...	110
	£458

Value of first crop, ...	350
	£108
Second year: contract for fencing another 100 acres, sowing it with wheat, reaping and thrashing, ...	150
Wages paid and horse keep for cultivating 100 acres of Indian corn, ...	150
	£408

His 200 acres of corn crop will now yield him from 600*l.* to 700*l.*, thus more than recompensing his outlay, and leaving plenty in hand to pay his first instalment, and to proceed with the vigorous cultivation of the land. The same sum which would be needed to start one son as a farmer of another man's high-rented land in England, would thus start four sons as the owners of farms, fenced, stocked, and under crop, on the fine prairie soils of Illinois.

I have in a previous letter pointed out the profitable nature of sheep farming in Illinois, and would again refer to it here, as an object well worthy of the consideration of young emigrant farmers. Merino sheep prove very healthy, and can be kept cheaply on the prairie. Their wool is nearly as valuable in America as in England, and the supply is not adequate to the increasing home consumption of that country. A large stock of sheep may be purchased with a small capital. I cannot help thinking, that the safest speculation for an enterprising immigrant farmer would be, the purchase of a section of land in the midst of untouched prairie, which he would enclose and crop, for the purpose of wintering a large sheep stock, which he might graze on the open prairie during the summer, at no other cost than that of herding.

The price of the land is the least consideration that a British emigrant need take into his calculations. For if he avails himself of the credit system, he may enter on as much or as little as he likes, for an immediate payment of only 3*s.* an acre; and his next payment is not due until he has been two years in possession, by which time the produce of the land ought to be much more than sufficient to meet it. The travelling expenses, the expense of maintenance, the building of a house, and the necessary outlays for stock, are nearly as great to start a small farm of forty acres as one of four times the extent. A man with his wife and four children, could not transport himself and them from this country to Illinois, and place himself comfortably even on a forty acre farm, for less than 100*l.* But 20*l.* more would suffice to place him in a farm of 160 acres. I cannot, therefore, advise men who are unable to scrape more together than will merely pay their travelling expenses to go to Illinois. And far less can I advise them to go farther west. Suppose they could obtain land in Iowa or Minnesota, 400 miles farther away, at only half the price, the saving of 1*s.* 6*d.* an acre in their deposit, would never compensate even the cost of travel for the additional distance, while every article which they require to purchase must bear an enhanced price from the same cause.

But there is one class of our labouring population, for whom Illinois offers great encouragement. Young men of intelligence and prudence, who have been brought up to agricultural pursuits, and are acquainted with the management of land and live stock, may do very well by hiring land from the owner. They get the farm fenced, stocked with necessary buildings, and with all requisites, except labour, for carrying it on. They furnish the labour, and share the produce with the owner. This is a transaction which is very common in Illinois; it answers the purpose of both parties; and a prudent active man who enters upon it, will generally in a few years realise enough to start himself in a farm of his own. I heard of many instances of great success attending this sort of arrangement, and from my personal knowledge, I am sure that there are many hundreds of our northern agricultural labourers possessed of the requisite skill and prudence, to ensure success. To such men, I should be happy to offer any information in my power, on application being made to me.

One great advantage that an emigrant of any class possesses in the Western States is, the facility with which land may be acquired; not merely its cheapness, but the readiness and simplicity with which it may be legally transferred. Every five pounds that a man saves, may at once be invested in land. He needs to run no risk of bank failures; and his landed investment is constantly improving in value, though he does nothing whatever with it. The same process is going on at home, but the labouring man at home cannot share in an advantage to which he himself contributes, as the land is too dear for him, and the cost of transferring a small parcel of it is nearly as great as the cost of the land itself. He is thus shut out altogether from the hope of ever being the possessor of land, and cannot therefore participate in that increasing value which is the good fortune of the rich alone.

Though I have doubts of the success of even prudent men who have no more capital than their wits, and no agricultural knowledge or experience, I feel bound to say that such is not the general opinion of experienced men in Illinois. One gentleman of high reputation and fortune in Chicago, assured me that he knew innumerable instances of people brought up in towns, with no knowledge of country life, and very limited means, who had blundered into experience and comfort in a year or two. In this country, he said, that every necessary of life is sluttishly plentiful; that it is not possible here to find a man hungry; that nature is so abundant that no prudent man can help becoming rich; and that, though they may sometimes have a set back for a time, they will soon rebound and take a fresh start,—and that there is plenty for all.

LETTER IX.

The Ague: Opinion of a leading Physician.—Easily curable.—Wisconsin.—Life-guardsman turned Implement Maker.—Success of Emigrants.—Madison.—Milwaukee.—Its Trade Facilities.—Schools.—Public Buildings.—Catholic Church.—Western Shore of Lake Michigan.—General Nature of Country.—Green Bay.—Early French Settlements.—Their Hold in the North-West.—Character of various Races of Settlers.—Price of Land.—Its Value dependent on cheap Access.—Interest of Money.—Credit low.—"Custom" pleaded for Abuses.—London Carpenter's Experience as a Settler.—The Mississippi.—Nails and Glass only allowed by American Government to their Surveying Engineers.—River Steamers.—Anecdotes.—Lake Pepin.—St. Croix.

I HAVE already referred to the risk of fever and ague which every emigrant to the western states must face. Before leaving Chicago I had an interview with a leading physician, who has been twenty years in practice in Illinois, and who kindly favoured me with his experience on this subject. He has found ague easily curable by proper management. It is usually preceded by diarrhœa and followed by fever, and the whole may be greatly mitigated, and even averted, by preventive and remedial treatment. It is caused in his opinion by excessive heat when it follows a moist summer. Summers of great heat, when the thermometer, for four or six weeks, is continuously 92°, 94°, or so in the shade, are always followed by an unhealthy season. Such seasons have been observed to come in succession at intervals of about ten years. For some years back there has been little sickness in the western states. But last year there was a great deal. In all new countries in these latitudes, especially after the ground is newly broken, he considers strangers more or less liable to it. But settlement, drainage, more comfortable houses, and care in diet and clothing gradually lessen its effects. In New York State, in the Genesee valley, about twenty-five years ago, people were extremely subject to ague. Now it is little known. Settlers coming from Europe are most liable to it when they come in summer. Arriving at the cool season in autumn, and going into a good house, with care not to expose themselves to hard labour in the sun, and to keep themselves well clothed, they are less likely to be attacked. Old people in his opinion ought not to come at all,—it is very fatal to them. Salts or purgative medicines taken at the commencement of the attack are very injurious, and he has found that the English emigrants are generally provided with such medicines, and take them at once, and without waiting for advice, on the plea that salts never did any body any harm. Nothing is more difficult than to remove the bad effects of this dose. The doctor said that Chicago was now almost free from ague, that typhus had taken its place in a greatly modified extent, and that pneumonia and rheumatism were the only other diseases that were severe.

Though I paid a second visit to Illinois before leaving the Western States I have now completed all that I mean to say upon it, and will ask the reader to accompany me into Wisconsin, and thence up the Mississippi to St. Paul's in Minnesota. The same railway facilities are found in the southern part of Wisconsin as in Illinois.

In the railway carriage I met a Sheffield man, who, though only thirty-five, has been nine years in the country, and had been previously for some years a soldier of the 1st Life Guards. He had been brought up a mechanic, started in Cincinnati as a maker of files when he first came out, and subsequently removed to this State as a partner in an agricultural implement factory. He has been doing a large business here, and a profitable one till this year, when the general failure of crops has dried up the source of his trade. He told me that the majority of the emigrants are people from towns who never saw a furrow turned till they came here, that they have everything to learn, great hardships to endure, but that with prudence and sobriety success is so certain that he had never seen an instance of it otherwise. He said, moreover, that such people, with dear-bought knowledge, turn out tolerable farmers in a short time.

The country towards Madison, the capital of the State, is dry prairie, lying on gravel for most of the way. The wheat crop here has been only half a crop of inferior quality, oats nearly a failure, but Indian corn good. Madison is prettily situated on a high ridge between two lakes.

But though this is the capital, Milwaukee, on Lake Michigan, is really the chief town of Wisconsin. It is one of the most picturesquely situated towns of any size that I have seen in the west. It is placed on both sides of a river which falls into a fine bay of Lake Michigan, the town rising from the valley of the river on either side to high bluffs facing the lake. The river is navigable from the lake, and vessels discharge and land their cargoes direct into, and from, the granaries and warehouses which line its banks. Tramways from the various lines of railroad run along the other sides of these warehouses, so that the greatest facilities are afforded for the transport and handling of produce and merchandise. The extent to which labour is economised in this way both here and at Chicago is really wonderful. By the aid of steam power half a million of bushels of grain can be daily received and shipped through the granaries of Chicago, the whole of it being weighed in draughts, of 400 bushels at a time, as it passes from the railroad to the vessel. This can be done at a cost of a farthing a bushel, and so quiet is the whole process that there is little external evidence of much business going on.

The finest church in Milwaukee is the Roman Catholic Cathedral, with the palace of the bishop on one side of it, and an orphan asylum on the other. There are many handsome private residences, some built of white marble, and the principal hotel of the city, the Newhall House, is very little inferior either in size, architecture, or interior fittings and arrangements to the Hotel de Louvre in Paris. There is a population of 60,000 in this city, which only twenty-three years ago was the site of a single log cabin, but which now, in the one month of October, ships a million bushels of wheat! From the bluffs the lake looks exactly like the sea, as no opposite shore can be seen, and the white-crested waves come rolling into the harbour just as they do on the Atlantic, though not with the same long and heavy swell. There are numerous schools in the city, free to all, and well endowed by the State. Education in the Western States seems to be far more highly prized than religion. I have often thought that the status of the schoolmaster and clergyman in Britain should be more nearly equal, but the latter ought not to be below the former, which is practically the case here. The Roman Catholics are certainly an exception, for wealth is lavished on their churches and cathedrals,—wealth drawn from the poor Irish, who, though they can't save money for themselves, are liberal to holy church. If one may judge from their edifices the Roman Catholics are making head in America, not so much in the way of converts, but because the Yankees and the Protestant settlers are too keen on business to pay sufficient attention to religion. The Catholics are moreover the only church in America which is bound together by an all pervading system of union.

From Milwaukee I proceeded by rail nearly 100 miles further to Lake Winnebago at Fond-du-lac. The country most of the way is timbered like Canada; indeed, all the western shore of Lake Michigan is so. There are oak openings here and there, which are little prairies of rather sandy soil in the midst of the woods, and which are all cultivated. The woodland is a reddish soil lying on gravelly hills, which is found to produce a good quality of wheat, the Milwaukee wheat holding a high character in the American market. The Indian corn seemed a small crop, and not much cultivated. After passing Horicon we enter on fine prairie, which continues for many miles through the county of Marquette, but all good land here, where improved, is, if anything, dearer than in Illinois. To the north of this the country is marshy or sandy, and, still farther north, it is all covered with pine timber,—the great timber region of Wisconsin, which yields more than 500 millions of feet of lumber in a year. Green Bay, which is a sheltered arm of Lake Michigan, 100 miles long and 30 broad, was the site of one of the earliest French settlements more than 200 years ago.

It is a remarkable fact, that before the pilgrim fathers landed at Plymouth, nay, even before the English cavaliers settled in Virginia, this far western country had been discovered by the French, who had sailed up the St. Lawrence and the lakes from Quebec, and whose missionaries, as early as 1624, were preaching the Gospel to the Indians on lakes Huron and Superior. By degrees, under the policy of Louis XIV., they stretched to the Mississippi, and were the first Europeans who held that magnificent valley from nearly its source to New Orleans. For a century they continued to hold the entire control of the North-West, till, in 1758, Wolfe's victory over Montcalm at Quebec wrested this dominion from the French. The poor French Canadians of Lower Canada are the only record of their power, and Green Bay, one seat of their first appearance in the west, is now being rapidly settled with Germans and Norwegians. From this point there has been a water connection formed with the Mississippi.

I was glad to learn that the Scotch settlers bear a high character in the west. They and the Germans are the most industrious, prudent, and successful. The Irish are met with everywhere, but generally improvident, hewers of wood and drawers of water, hanging on about the towns, doing all the dirty and hard work, and the menial offices at the hotels. The English or Scotch are never met with in such situations, the Germans often, the Irish and Negroes always.

There is abundance of land to be had in Wisconsin. I was offered 6000 acres of prairie, 300 miles west of Lake Michigan, on a river which flows into the Mississippi at Lake Pepin, for 12*s.* an acre; and woodland, near the north-western shore of Lake Michigan, may be had at the same price. Farther north and west, there are thousands of acres to be had at 5*s.* an acre. But these out of the way places are only seemingly cheap. The cost of transport, so long as there are no railways, renders the crops of the settler valueless. A man will make a better bargain who pays 50*s.* an acre for his land, when he can sell his first crop for much more than the difference of price.

The interest of money lent on mortgage of real estate, is from 12 to 18 per cent. There is no doubt the risk of having the property thrown on one's hands, but the legitimate profits in this vast new country afford a very high rate of interest. The credit of the State has been also somewhat damaged by the shameful conduct of its public men. I had heard rumours of this before, but did not believe it till it was proved to me on the spot. The governor of the time, and the majority of both branches of the State legislature, were convicted, on their own confession before a committee of investigation, of having been systematically bribed by a railway company, who thereby secured a vast grant of public land! The governor got 20,000*l.*, each Senator 2,000*l.*, and each member of the Lower House 1000*l.*, luckily in railway shares, which at present are of very little value. Who was the governor? A man of no note, purposely chosen for his obscurity, because obnoxious to neither political party; for whenever parties are nearly balanced they seek out for the highest office men of no prominence, as the really able men have offended one side or the other.

It is amusing, when one is used to it, to see the cool way in which the public in these western States are trampled upon by officials, and the servants of steam-boats and railway companies, and how quietly they submit to it. I had taken a through ticket like other passengers, "from Fond du Lac to Chicago." Such were the terms on the ticket, and of course that was our contract with the company. We reached a town, and were told we must get into and pay for an omnibus to convey us between the stations. We proceeded. One poor delicate German wished to be put down at an intermediate station, on another line, for which he had taken his through ticket. The bus drew up, dropped him ankle deep in mud in the middle of the road; the driver never left his seat, but told him to catch his trunk, a very large one, trundled it off the roof, nearly upsetting the poor German, and the last we saw of him he was hauling the trunk through the mud to the station. Not a porter offered the slightest aid. I indignantly remarked to the rest of the passengers that this was shameful, and expressed surprise that a free people would submit to such insolent imposition. I said that in my country the railway company would be bound by law to fulfil their contract, and to provide at their own cost the necessary transport from station to station. An American answered that it was certainly very bad, but "guessed that the company could defend themselves at law by the plea of 'custom.'" I expressed astonishment, when a 'cute Yankee beside me interposed, "I reckon, stranger, that you are about right. I don't think custom would carry it. Least I knew a case where it didn't. Two steamers were racing on the Mississippi. A passenger was seen,—both made for him. A plank was shot ashore from the foremost boat and he stepped on to it. But they were in such a hurry that they drew in the plank and threw him into the river. An action was raised, and the owners of the boat pleaded 'custom.' But the judge held that a contract to 'carry' could not be fulfilled by throwing a man into the water, notwithstanding custom." "Was the man drowned?" said I. "No, but he war darn't near't."

On my way across the State to Prairie du Chien I met a Cornishman, who had worked for many years in London at his trade as a carpenter, and who had come out to Wisconsin and bought eighty acres of "oak opening." His first crop of wheat yielded twenty bushels an acre; and the second, notwithstanding the unfavourable season, fifteen bushels of spring wheat, which he was then selling at 3*s.* 9*d.* a bushel. Besides looking after his farm, he works at his trade, at which he earns 5*s.* a week more than he used to make in England. But a tradesman without land will, in his opinion, spend all the difference in keeping himself. Men with families, he said, could get on better here than at home, provided they can buy land enough to support their families, and when they save any money they can buy more land, an impossibility at home. He thinks that agricultural labourers would benefit by a change to the West more than any other class, because their wages at home are so low, and their peculiar skill is the thing most needed in an agricultural country.

At Prairie du Chien we found ourselves on the Mississippi, the Father of Waters, which at this point is nearly 2000 miles from the sea. This was formerly a French trading post, and more recently a frontier post of the Americans. There is here a loop-holed barrack capable of lodging 400 men. It was discontinued soon after the last fight with the Indians here in 1833, the frontier garrison having since that time moved many hundred miles farther west. It is now abandoned and going to ruin. The American system, as I was informed by an engineer officer of the service, is to spend no money in keeping up establishments after the object has been accomplished. Their surveying officers on the frontier are allowed only nails and glass, and with these they may erect quarters if they like. If not, they may live in their tents,—at all events, they are not permitted to spend public money.

We here embarked on the Mississippi for St. Paul's, a voyage up the river of 300 miles. The river at this place is about as wide as the Rhine at Cologne, but with a less rapid current and not so deep a stream. The west side is very picturesque; a series of limestone-bluffs, 200 feet high, covered on their face and summit with autumn tinted woods, and broken into irregular forms by little valleys branching off from the main stream. As we quietly proceed on our course, every new reach opens out a fresh scene of beauty, and we are soon shut in on both sides by lofty ridges of limestone rock. In many places this ridge retires a short way from the water, its sharp edge disappears, and a round grassy face, smooth and regular as a lawn, runs up within twenty feet of the top of the sharp peak or frowning rock which crowns the whole. Single trees are scattered like ornamental timber over the green hill sides, which presents the most charming natural sites for building. But houses there are none, except here and there at a landing-place on the river, where a wooden store and "office" invite the traveller to land and become an unit in the incipient "city." There are also huts on the edge of the water at convenient points for "wooding," occupied by wood-cutters, who prepare fuel for the steam-vessels. In summer the banks are infested with musquitoes, and the people live in the open air, round large fires, to protect themselves from the insect. The farther north you go in these latitudes, during the short but hot summer, the more you are liable to be tormented by musquitoes.

Our steamer, which draws only 28 inches of water, is a huge structure. The saloon is 200 feet long and 8 feet high, with Gothic roof painted white and gold. There are little sleeping cabins along both sides of the entire length, sufficient to accommodate 130 passengers. The fare includes provisions, and an abundant table is served three times a day. There is hardly any difference in the meals in this western country, except that to breakfast and supper we are offered tea and coffee, while at dinner cold water is the only beverage. The manners of the people we meet with on the Mississippi are not a whit exaggerated by Dickens in "Martin Chuzzlewit." I have met with instances of every dirty habit which he describes, and any Englishman who desires to see the West must steel himself against disagreeable incidents of hourly occurrence. He will be amply repaid for any inconvenience of this kind by the vastness of the new scenes constantly opening before him.

The high-pressure engine every minute emits a melancholy sigh, but it drives us quietly along against the stream at a tolerable rate. A traveller on the shore holds up his umbrella; the huge vessel, as if watching him, sheers in towards the soft bank, runs her nose upon it, a plank is shoved out, and the "gent" walks on board: we back off and proceed. The ship is managed by a pilot, who has a glass-house elevated between the paddle-boxes, above the whole superstructure, and from this commanding position both steers the vessel and communicates his orders to the engineer by a signal bell. The captain's business seems to be to keep order in the ship, and to take the head of the table at the various meals. While standing beside him at the front of the elevated deck, our ship began to back in the middle of the river. "What's the matter?" said I. "Smelt a bar, I guess," said the captain, "and backing out in time." "These snags," said I, pointing to one, "must be troublesome at night; don't they take them out?" "No, we know about where they *air*, and keep pretty clear of them." "Any other danger on the river, Captain?" "No, only two, a sink or a burn up. We sometimes snag—not often—and sometimes go afire. But we seldom have an accident. In June last there was a burn up,—a few miles below;—seldom happens though!" "You have no heavy sea to trouble you, at any rate;" I threw in by way of comfort. "I guess not," said the captain, turning full upon me, "we build these steamers strong enough for their purpose, Mister, and as light as possible. They are for the river, not the lake. In a heavy sea they would double up in a clip; and that's a fact."

There are many sand-bars in the river, which, at this season are so near the surface that the vessel sometimes sticks. But in the bows two great legs or stilts are fixed, like little masts, with blocks and tackle, and when the steamer "bars," down go the stilts, the tackle is made fast to the capstan, the men pull upon it, and raise the ship a foot or more, clean off the bottom, at the bows. The paddles are then set on full steam, and the vessel is literally jumped over the bar. I asked the captain if by this means he could get over any ordinary bar. "I reckon I could lift her over the river bank, if she would hang together," was his reply.

We passed at night through Lake Pepin, an expansion of the river, from two to three miles broad and twenty-five miles long. The scenery is said to be very beautiful, but we could only catch a glimpse of the "Maiden's Rock,' which rises sheer up about 200 feet from the water's edge. A romantic American, an individal rarely met with, told me its story in the starlight. Winona, the daughter of a celebrated Indian warrior, had been betrothed by her father against her wish. The wedding-day was appointed, and the feast was being prepared. She and her young companions went out to gather a berry that grows among the rocks. It was a summer evening, and, busied in their occupation, the rest did not observe that Winona had parted from them. Suddenly from the summit of the rock a low cry was heard. It was the death song of Winona, who, in a moment more, with one spring from the edge of the precipice, buried herself in the lake.

At a place called Prescott, at the junction of the St. Croix river, on the Wisconsin side, the country is remarkably pretty, wooded and park like, with rounded grassy knolls 100 feet high, which slope down towards the water, terminating in a precipitous limestone bank. Here, in some places, the prairie falls gradually to the edge of the river. The country for some distance back is all bought and occupied, but none of the settlers seem to choose a residence among the wooded glens. A few hours more bring us to St. Paul's.

LETTER X.

St. Paul's.—Route to Red River.—Minnesota.—Daily Newspapers.—Market Place.—Red Indians selling wild Ducks.—American Militia.—Fort Snelling.—Minnehaha.—Falls of St. Anthony.—Lands and Funds set apart for Public Objects.—The Credit System.—Down the River to Dubuque.—Burlington.—Iowa.—Natural Obstacle to Progress of Population Westwards.—Wages.—Nauvoo.—St. Louis in Missouri.—Slave State.—Iron Mountain.—Relative Cost of Production of British and American Iron.

St. Paul's, the capital of Minnesota, the last State admitted to the Union, stands very beautifully on a sloping limestone ridge of the Mississippi, upwards of 2000 miles from its mouth at New Orleans. It may be regarded as the head of uninterrupted navigation, for the Falls of St. Anthony, only nine miles further up, close the passage. Above the Falls, however, steamers ply 150 miles still further northwest. From this highest point it is proposed to make a land connection with the Red River, which flows north, and is navigable for 300 miles before it enters the British territory at Pembina.

The new State of Minnesota has an area considerably greater than the British Isles. The southern part is chiefly prairie, very level for great distances west, as was shown to me in a section of the railway now being constructed. The soil is considerably more sandy than that of Illinois; the winters are intensely cold, but the summers, though comparatively short, generally mature the various corn crops which are cultivated. This State has its northern boundary along the British territory, at present possessed by the Hudson's Bay Company. The crops for the two last years have been threatened, and partially injured, by a plague of grasshoppers.

A bridge is in course of construction to connect the two banks of the river, on both of which the city is being built, the one to be called East, and the other West St. Paul's. The houses are solidly built of limestone, the material for the walls and mortar being found in excavating the foundation. All kinds of public improvement are rapidly carried into execution, city bonds being issued to defray the cost. These bonds can be purchased to yield 12 or 15 per cent., and may prove a good security if population continues to flock to Minnesota. Banks, land agency offices, and newspapers, are already numerous in the city. The newspapers are dailies, and in walking out early on the Sunday morning, before many people were stirring, I was surprised to observe the morning papers already laid on the handle of every door, or shoved in below it, ready for the owner's perusal as soon as he should make his appearance.

The market-place on Saturday was thronged with people buying and selling their various produce. The Irish had sacks of potatoes and other vegetables, the Americans dealt in beef; but the most remarkable looking merchants there were the Red Indians, who were selling wild ducks. They were in considerable numbers, both men and women, in their native costume, the men generally carrying good double-barrelled guns. There are still several Indian tribes at no great distance from St. Paul's, who live entirely by the produce of their guns.

I was introduced here to a State senator who, with basket on his arm, was making his morning's market. I was afterwards indebted to this gentleman for a presentation to the governor, whilst he was reviewing a corps of volunteers at the State-house. They were very soldierly-like men, their uniform more like the French than English, and they seemed to go through their evolutions very creditably. They had a brass gun, and three or four artillery men in the corps. The United States have an enrolled and organised militia of upwards of two millions and a half, from which a very formidable army might readily be selected. The people are fond of soldiering. In every considerable town, some volunteer cavalry or infantry corps will be found parading about, but I never saw a soldier of the regular army all the time I was in the Union. These are all posted in the interior of the continent on the Indian frontier.

In pursuing our course to the Falls of St. Anthony, we skirt along between the prairie country and the bank of the Mississippi. We cross the river by a ferry below Fort Snelling, one of the old frontier posts now abandoned. It stands on the point of the promontory, which juts out into the junction of the Minnesota River with the Mississippi, at an elevation of 150 feet, and must have been capable of easy defence against any sort of Indian warfare. Two miles farther we came to a little gushing stream, where is laid the most beautiful scene of Longfellow's Indian poem, "Hiawatha,"

> "Where the falls of Minnehaha
> Flash and gleam in shining reaches,
> Leap and laugh among the woodlands."

We dived into the little glen, admired the waterfall, drank of its fresh waters, and finally cut walking-sticks in remembrance of it. It is certainly pretty; but, as an American has described it, unusually "neat." The water pours over a rock through a groove which exactly fits it, and it runs away below with all the regularity of a mill stream. Nevertheless, for the lovely Minnehaha's sake, we did our best to admire it. There is a great distinction between American and European scenery in this, that in America there are few accessories to the scene.

There is the waterfall, but no enclosing mountain—no dashing along over rocky bed before the final leap, and but a very tame gorge below. The face of the country is generally monotonous,—hundreds of miles of bare prairie, breaking down at its edges into natural troughs for the water. The traveller over the prairie comes upon the great river suddenly; and sees, perhaps a hundred feet below him, the vast stream flowing along the hollow trough which it has worn for itself in the course of ages.

A few miles farther brought us to St. Anthony, where the Mississippi makes a leap over the rocks of some twenty or thirty feet. The river was low; and as we were then nearly the whole length of Europe from its mouth, we did not expect too much. Moreover, both sides of it belong to Jonathan, and he is a deal too sharp to throw away so good a mill power. On each side, then, the main body of the water is caught, and turned to the servile purpose of sawing lumber. The surplus water is left to run off in the centre, where it forms a little green imitation of the Great Horse Shoe Fall of Niagara. We literally "hunted" this waterfall, for we were a good hour jumping across the floating logs, and along the various dam faces, before we reached the best point of view. The young American lumbermen employed here are fine stalwart men, extremely expert in the use of the axe, by which they earn several dollars a-day at piecework.

We returned by the other side of the river, which is crossed by a suspension bridge above the Falls. This brought us to the new city of St. Anthony and its vast hotel, now seemingly empty, then past a college which has recently been built for higher class education.

The eighteenth part of all the public lands in Minnesota is set apart for the support of schools; 46,000 acres more are appropriated for a State university; 6400 acres for the erection of public buildings at the seat of government, and 5 per cent. of the sales of all public lands are granted to the State, by Congress, for the construction of public roads and internal improvements. Besides this, every alternate square mile of land, for five miles on each side of the lines, is granted to aid in the construction of various lines of railway which are intended to traverse the State.

The system of credit established throughout the American Union, if very unsound in all times of difficulty, is certainly productive of many useful results. For instance, the State of Minnesota has fine unoccupied land, which is of no value so long as it remains inaccessible. It is determined to construct a railway, and the State finds the funds in this manner:—it issues bonds bearing 6 or 7 per cent. interest, which are handed over to the contractor as his work progresses. These bonds may not be very saleable out of the State, but the contractor lodges them with the State treasurer, and obtains, in lieu, 90 per cent. of their amount in authorized notes of issue. With these he pays his wages and bills, finishes another section of road, receives a second instalment of State stock, makes a second issue of notes, and so the thing goes on until the road is made, the country opened up, and produce brought to market. The bonds are cleared off as the land is sold, and everybody is benefited.

There is yet only one way of going to or returning from St. Paul's, and we therefore took steamer down the Mississippi over our former course as far as Prarie du Chien. We proceeded about 100 miles farther to Dubuque, in Iowa, where, and at Burlington, 200 miles farther down, I had an opportunity of seeing some of the prairie lands of that State. An Irish settler, who had been seven years in the neighbourhood of Dubuque, was enthusiastic in his praise of the country. He said that the grass was as good as in Ireland, the crops of corn better, and more wealth of tin and lead in the bowels of the earth than would make the world rich. He told me of three cousins of his who came out three years ago, with money enough to buy 600 acres among them;—that their cattle, horses, and swine had increased so much already that they could not count them, and that they had been lately offered twice as much for 100 acres of their land as they had originally paid for the whole! There is a rich mineral district round Dubuque, similar in its character to that of Galena, on the opposite shore of the river, in Illinois.

Burlington is finely placed on the side of a bluff rising from the river, to a height of 150 feet. It has a population of 15,000, and is likely to increase, as the system of railways which centre in it are opened up westwards. It is the chief town of the southern part of Iowa, and the railway now being constructed from it to the Missouri River, traverses the finest portion of the State. By the month of May, this line is expected to be opened as far as the Des Moines River. I travelled on it to Fairfield, its present terminus, through a country of prairie and woodland intermixed. It seemed not more rolling than Illinois, and not so rich. Very little clover or blue grass is seen growing along the line, such as cheers the eye to the westward of Mendota. The buildings in the towns are inferior, and the country generally looks poorer. There is no land for sale along the line, at government prices, nearer than fifteen miles distant,—and enclosed land within that distance sells at from ten to twenty dollars (2*l.* to 4*l.*) an acre. But the market on the Mississippi is not equal to that of Chicago, and the land of Iowa cannot be so profitable as that of Illinois, for the cost of the additional transit must always operate against the former. Beyond Fairfield, the line traverses a coal country, and its point of junction with the Missouri River is opposite to the terminus of the Platte River, which is said to be navigable 600 miles farther west, so boundless is the extent of this country. Here is a railway being constructed for nearly 300 miles west of the Mississippi, through a region the greater part of which has yet been trodden only by the Indian and the government surveyor, and yet its terminus is but the starting point for another 600 miles, through a country as extensive as Great Britain, on the confines of which are the newly discovered gold fields of Kansas and Nebraska.

There is ground, however, for believing that, beyond a certain point westwards, the country will be found from climate not so well adapted to the maintenance of population. The change begins at the meridian of 95° west longitude. The air then assumes an aridity not found anywhere to the east of it, and at the 98th meridian it presents an abrupt contrast with the country east. Mr. Blodgett, in his very able and interesting work on the climate of North America, points out this fact, with the explanation that the plains here have an elevation of 2000 feet on an average. This arid climate is not only unfavorable to the culture of corn and grass, but is probably productive of those enormous flights of grasshoppers which for two successive seasons have seriously damaged the crops of Iowa and Minnesota. The western side of both these States is within the influence of this aridity of atmosphere, as is likewise the British settlement on the Red River, where on several occasions the crops have been utterly destroyed by grasshoppers.

There is a population of 600,000 in Iowa. The foreigners are chiefly Germans and Irish, the latter mostly railway labourers. Wages are now a dollar a-day.

We again embarked on the Mississippi, passing Nauvoo, the first Mormon settlement in America, now broken up some years ago. After a farther voyage of 200 miles we reached St. Louis, one of the great cities of America, with a population of about 130,000. It was formerly called the "Mound" city, from a great mound at the base of which it was first settled by the Spaniards, and which is said by the Indians to have been the burial place of their fathers for ages,—by others to have been formed by the Aztecs before the time of the Indians,—and not improbably formed by nature before either one or other. Jonathan however has no respect for antiquity, and this famous "mound" will soon be entirely cut away to fill up an embankment on the river, and to enable certain streets to be laid out in a straight line. About the half of it is already gone. In another year probably not a vestige will remain of that which gave the name at first to this city, which in all ages has been a great landmark on the river, the sacred place of the Indians, and which, with a little veneration and good taste, might have been left to form a great feature of attraction and interest in this splendid city.

St. Louis is a fine old town for America, situated on rising ground, and covering a large extent of surface,—with handsome streets, stores, shops, squares, hotels, and churches. The quays along the river are literally crowded with steamers, lying abreast of each other in tiers three and four deep. This is a slave State, the domestics are nearly all blacks, the property of their masters.

The richest part of the country is along the banks of the river Missouri. Tobacco is chiefly cultivated there. The southern part of the State is rocky and wooded. Very little of it has been taken up by settlers, and any quantity may be purchased at little more than 6*d.* an acre. But it is not agricultural land, is very inaccessible, and Europeans are not fond of settling in a slave State.

There is a famous deposit of iron in this State, called the Iron Mountain, which we were anxious to see. There is a railway to it, eighty miles in length from St. Louis. The line runs along the Mississippi for twenty-four miles, at the base of limestone bluffs which are covered with oak, and by an undergrowth of the beautiful red-leaved sumac. The banks and sweeps of the river are very picturesque. When we leave the river, we pass over an undulating wooded country with many streams. Cleared spots yield good crops of corn, now in shock, and the young wheat is fresh and green. But the land generally is not very tempting for agricultural purposes. Vine culture, however, has been successfully introduced by the German settlers, and is making progress.

The Iron Mountain covers a surface of 500 acres; it rises to a height of between 200 and 300 feet above its base, and nearly 1000 above the bed of the Mississippi, from which it is thirty-eight miles distant. It is clothed with young oak and hickory trees, which thrive well on the scanty soil between the crevices of the rock, and among the loose stones with which it is covered. The whole mountain, and every stone upon it, is nearly pure iron ore, there being only 2 to 3 per cent. of silica and alumina, 65 per cent. of pure iron, and the remainder oxygen. The quantity above the base is estimated at two million tons. A bore of 150 feet has been made, at the base, without meeting with any change, and the geologist of the State says that it may be of any depth, if it is, as he supposes, a mass of iron which has been upheaved through a fissure of the earth's crust.

No coal is found nearer than that of Illinois, opposite to St. Louis, and the cost of carriage thus operates as a great bar to the profitable working of this deposit. The Glasgow blackband iron is found in combination with sufficient carbon to calcine, and, in some cases, even to smelt it; and there are always strata of coal found with the iron in Scotland, which gives a great advantage to the Scotch iron-masters. Lime, which is used as a flux, is got in close proximity to this mountain.

Wood charcoal is used in smelting; but this is found expensive, and a tramway is now being made from the quarry to the railway, with the view of transporting the iron stone to St. Louis, in the neighbourhood of coal. When the ore can be loaded directly from the rock, it may be placed on the cars at the cost of 8*d.* or 9*d.* a ton, and may be delivered at St. Louis at a cost of about 4*s.*

There is another iron mountain on Lake Superior, of equally good quality and purity. An axle made of it was said to have stood the following comparative test: blows before breaking, Lake Superior iron, 177: best Swedish, 77: English Low Moor, 46. The present cost of producing one ton of this pig iron on the wharf is $18; the cost of the St. Louis, on the wharf, is $16.

At Pittsburg and Wheeling, where the principal iron works of America are situated, they have the same natural advantages for the economical production of the metal as in England, and the supply is inexhaustible. And yet the United States pay more than 5,000,000*l.* a year for foreign iron. The native manufacturer is protected by an import duty of 24 per cent., and freight and charges will amount to 10 per cent. more. And yet with this disadvantage of 34 per cent. against him, the English iron-master can undersell the American at his own door! The Americans are careless in their management, they have not sufficient capital embarked in their works, and they expect larger profits than the English are contented with. There is not much difference in the actual cost of labour: it may be a trifle higher in America: but an American capitalist cannot obtain the same constant supply of skilled labour as the Englishman can command. The country is so vast, and the temptation to other and easier pursuits so great, that there is no constancy to certain employments as in England. The labouring population in America is not stable, it is a shifting, unsteady, improving, advancing mass. And, for many years to come, an old populous country like England will continue to have the advantage of her, in every kind of work which demands concentration of power, and the application of large individual capital.

But in some States great progress is being made in the manufacture of iron. Between 1850 and 1857, the quantity of pig iron produced in Ohio increased from 52,000 tons to 106,000 tons, though even this quantity is little more than one-third the annual produce of the iron-works of a single great firm in Scotland

LETTER XI.

The Ohio.—Cincinnati.—Vine Culture.—Kentucky.—Bourbon Whisky.—Cincinnati to Columbus.—Small Farmers decreasing.—Westward Movement of Population.—Indian Corn never fails.—Wheat precarious.—Average Produce very low.—Live Stock.—Profits of Farming.—Labour economised by Steam.—Quantity of Whisky produced.—Public Expenditure on Education.—Compared with England.—Rate of Taxation.—Crossing the Alleghanies.—Virginia.—Maryland.—Washington.—Baltimore.—Philadelphia.—New York.—Boston.—Home.

I PASSED rapidly during the night through Indiana, reaching the Ohio River at Lawrenceburg the next morning. The banks on both sides of this muddy stream are high and clothed with wood. They recede some distance from the river, leaving a valley which is extremely rich and highly cultivated. Near Cincinnati the hill sides are covered with vineyards, with pretty rural villas interspersed. Vine culture has been profitable here, and is extending. Ohio produces annually 700,000 gallons of Catawba champagne, a sweet sparkling wine, not equal to that of France.

Forty years ago Cincinnati was a small town. It has now a population of 200,000, and is a very fine city, with many public buildings, shops, and streets of great architectural elegance. There is an excllent public library and news-room, to which strangers are admitted.

I took a run by railway 100 miles into Kentucky to Lexington, the home of Henry Clay. This State is famous for its grazing properties, a red soil lying on limestone, and the fields, where the country is cleared of timber, are green and rich-looking. The cattle are well-bred short-horns. Very large herds of mules may be seen penned together, 100 or more in one lot. No stock pays better. Pigs are likewise fed in large numbers together, penned on a cleared piece of Indian corn, with which they are regularly and abundantly supplied. In Bourbon county a famous whisky is made from Indian corn, which bears as high a character in the States as Islay or Glenlivat does in Scotland. This is a slave State, and the field labourers are mostly slaves.

After returning to Cincinnati we pursued our journey eastwards, passing for some miles through the pretty valley of the Little Miami. The rest of the country to Columbus, the capital of the State, is partially cleared, but chiefly covered with primeval forest. An American in the railway carriage was complaining that this State was rapidly going into the hands of large landholders, and that the poor man would soon have no place. But the poor men are voluntary emigrants. They find it a thankless task to struggle with the forest, sell their farms to the adjoining proprietors, and push off westwards to the rich open prairies. Those who remain gradually absorb the neighbouring lands, and the uncleared country continues undisturbed. Nor will it be much disturbed for a generation or two, for there is lake and railroad access to the prairies, and men will not toil at these wooded solitudes when they can turn their furrow without impediment on the black prairie.

In the Ohio State Report for 1857, this movement of the agricultural population is referred to at some length, and it is there shown that a decrease in numbers had taken place in many counties, and that in five townships named, in one of the best wheat counties of the State, the farming population had decreased 6 per cent. through the emigration of "small farmers seeking a better home on the virgin soil of the West." It is there stated that not less than 140,000 persons had thus emigrated from Ohio since 1850. We cannot feel much surprised, therefore, that the Indian corn crop of Ohio in 1856 was a million bushels less than that of 1849, and the wheat crop of 1856 actually a million bushels less than it was seventeen years before, in 1839.

From the results of a series of crops extending over nine years, it may be inferred that Indian corn never fails in Ohio, but that the produce alternates in alternate years; that the years of least productiveness are those of more than usual dryness; and that the difference between a good and a bad season may make a difference of half the produce. Between the crop of 1855 and that of 1856 there was a difference of 30,000,000 bushels.

The production of wheat in Ohio has been diminishing during the last eight years, though this received a slight check during the three years of high prices before 1858. It is, however, quite clear that wheat is found to be a very precarious crop in Ohio. While Indian corn is rarely injured by anything but drought, wheat has to contend not only with that, but with other enemies not less destructive, viz. winter freezing, rust, the midge, and red weevil. The average produce has thereby been reduced to fourteen bushels an acre, and the average for the State in 1854 was only eight bushels! Such scanty crops can leave no profit to the farmer.

Nor have animal products undergone any material increase during the last eight years. Horses and mules have increased considerably, but the increase of cattle is met by a proportional decrease of sheep, and the numbers of swine, in the production of which Ohio is the chief State of the Union, have not materially augmented during the last seventeen years. In 1840 the number was 2,099,000, and in 1857 2,331,000.

The net income received from cultivated land in Ohio is reckoned at 10 per cent. This includes rent and profit, except that portion of the produce which is consumed in the maintenance of the farmer's family, and which is included in the cost of production. The cost of farm labour averages one dollar a day, without board.

The extent to which labour is economised in this State by steam power is estimated to be equal to the labour of 700,000 men. The number of reaping and mowing machines manufactured in 1857 was about 7000, which is seven times as many as in all England for either that year or last. It is by means like these that America accomplishes, with comparatively few hands, the cultivation of such vast extents of land.

The quantity of whisky produced in Ohio is extraordinary, being not less than 26,000,000 of gallons annually. This is double the quantity made in all Scotland, and very nearly equals the annual consumption of spirits in the whole of the United Kingdom. The quantity of spirits consumed in Great Britain has been computed to be equal to one gallon, or six bottles, a head for every person of all ages and sexes of the entire population. But Ohio leaves us far in the distance, as her production of whisky is equal to sixty-eight bottles for every individual of her entire population. No argument against the temperate habits of the people can, however, be founded on this fact with any certainty, as a very large quantity of spirits is exported to other States, and much of it is used in the arts and manufactures of the country.

If pre-eminent in the production of whisky, Ohio stands high in the support of education. She expends half a million sterling yearly in the maintenance of schools, which is at the rate of 12*s.* 6*d.* for each child in the State. The public expenditure in Great Britain for the same object does not exceed 2*s.* for each child in the population. It is a very remarkable fact that this single State, which, in the beginning of this century was an almost untrodden forest, spent nearly as much of the public taxes in 1856 on the support of education as our own old, populous, and wealthy country. Nor is this the whole property devoted to education; for, besides the amount which I have mentioned as levied by taxation, the thirty-sixth part of all the lands, amounting to 800,000 acres, are set aside for the support of schools.

The average rate of taxation for Ohio is 1·02 per cent. on the estimated capital value of the entire property of every kind in the State. As the net income from land is estimated at 10 per cent., this rate of taxation on capital is equivalent to a little over 10 per cent. on the annual income. And yet this State is reckoned to be moderately taxed compared with many others.

After visiting Columbus, the capital of the State, we pushed on to Wheeling, where we crossed the Ohio river, and entered Virginia. Day was just breaking as we reached Altamont, the loftiest station on the Alleghanies, and we then commenced our descent on the eastern side of the mountains. They are clothed with wood to their summits, and the glens along which the train shot rapidly out and in are pretty, but not wild. We followed the course of a wimpling streamlet which gathered to itself all the little rills from every hill side, and during the ten hours of our continuous journey, saw it growing by degrees, in a course of 150 miles, into a majestic river, the Potomac. The scenery was very interesting during the whole day's ride, but became extremely fine as we approached Harper's Ferry, where there is but one gorge through the mountains for river, canal, and railroad. The mountains rise on either side to a considerable height, the river dashes along against the precipitous rock on both sides, and the railroad for a mile or more is carried on piers above the stream till the gorge is passed, and a narrow strip of ground gives it a firmer footing. We soon reach the low country, in Maryland, which begins to resemble England. The farm-houses are substantial, the fields often enclosed with thorn hedges, the wheat and sown grass a lively green. After traversing some extent of woodland country we reached Washington in the evening.

This city, which is the seat of government, has been laid out on a scale of magnificence, the first outline of which will not soon, if ever, be filled up. It is not a place where business is likely to centre, and the President has no court to attract the residence of the wealthy. Thus, except when Congress is in session, there are few people of consideration in the city, except the President and his ministers, the heads of departments, and foreign ambassadors. Standing on the dome of the capitol the fine broad streets are seen radiating in straight lines, but all terminating abruptly in the primeval forest. The capitol is a building of great magnificence, two wings of white marble having recently been added to it. The two branches of the national legislature occupy each wing, and ornament is very lavishly bestowed in the decoration of the interior of both. Each member has a desk in front of his seat, with his name affixed to it, the seats being drawn for by lot at the commencement of every new Congress. The Smithsonian Institute, the Observatory, and the Patent Office, are the three other public buildings in Washington most worthy of inspection.

The country between Washington and Baltimore is undulating and wooded, but the soil all the way is either a wet clay or sand,—a poor country for man or beast.

The city of Baltimore, with a population of 170,000, lies low* on a bay of the Chesapeake. The whole country here is indented with bays, which are both picturesque and conven-

* Baltimore is built on several hills. The business part only is along the river—most of the city being elevated.—AM. ED.

ient. The chief business on the quays seemed to be the loading and unloading of tobacco. There is a fine monumental pillar to Washington in the upper part of the town, and many large and handsome residences in that neighbourhood. But it was not thought safe to walk alone in the streets after dark. A system of terrorism had been established by the "Know-nothing," or American party, in the city, which had led to several murders, and law and order were being set at defiance. The mayor had expressed a determination to put this down, but it seemed questionable whether he possessed the power.

The ride from Baltimore to Philadelphia is very pleasing. We pass numerous bays of the Chesapeake, until we cross the Susquehanna below Harford,—and, on reaching the banks of the Delaware, the country becomes fertile and cultivated. The position of Philadelphia is remarkably fine, placed on the apex of land between the Delaware and Schuylkill. A great city with half a million of inhabitants, it seemed designed by nature to be the capital of the country. There is more of the staidness, and quiet respectability, of an old and wealthy country about this city and its people, than I have seen in any other part of the United States. Many of the shops in the principal streets are built of white marble, and one or two of them exceed in elegance of external appearance any that I have ever seen either in London or Paris.

After a short stay we continued our journey to New York and Boston, from which place I embarked in the good steam ship "America," under the careful guidance of Captain Millar, and, after a pleasant, though somewhat stormy passage of ten days, arrived again in safety in Liverpool.

APPENDICES.

APPENDIX I.—COMPOSITION OF PRAIRIE AND OTHER FERTILE SOILS.

	Prairie Soils from Illinois Central Railway Lands.				Soil from Old Red Sandstone of England.	Wheat Soils from Scotland.			
	Analysed by Professor Voelcker, Consulting Chemist of the Royal Agricultural Society of England.					Anlaysed by Professor Anderson, Chemist to the Highland Agricultural Society of Scotland.			
	No. I.	No. II.	No. III.	No. IV.		Mid Lothian.	East Lothian.	Perthshire.	Berwickshire.
Organic Matter and Water of combination	7·54	5·76	9·77	9·05	4·38	10·19	6·32	8·55	6·67
Alumina	6·67	6·55	8·58	8·74	4·51	6·93	5·54	14·04	7·36
Oxides of Iron	4·95	2·57	4·13	4·30	3·15	5·17	4·41	4·87	4·32
Lime	1·37	·35	1·84	1·13	·77	1·22	1·39	0·83	2·70
Magnesia	1·03	1·53	·82	·61	·63	1·08	0·74	1·02	1·63
Potash	1·69	1·40	1·20	1·29	·74	0·35	1·71	2·80	0·55
Soda	·82	·53	·83	·50	·22	0·43	0·67	1·43	0·36
Phosphoric Acid	·08	·05	·12	·10	·12	0·43	0·14	0·24	0·22
Sulphuric Acid	·07	·05	·14	·08	·06	0·04	0·10	0·09	0·05
Silica	75·04	80·68	71·75	74·11	85·11	71·55	74·39	63·19	73·52
Water	....	....	....	....	....	2·58	4·42	2·70	2·50
Carbonic Acid and loss	·74	·53	·82	·09	·31	0·03	0·17	0·05	·12
	100·00	100·00	100·00	100·00	100·00	100·00	100·00	100·00	100·00
Containing Nitrogen	·30	·26	·33	·34	·18	22	13	·21	·14
Equal to Ammonia	·36	·31	·40	·41	·12				
MECHANICAL ANALYSIS.									
Clay	64·14	46·76	58·90	62·75	18·09				
Lime	1·37	·35	1·84	1·13	1·37				
Sand	26·95	47·13	29·49	27·07	76·16				
Organic Matter	7·54	5·76	9·77	9·05	4·38				
	100·00	100·00	100·00	100·00	100·00				

Appendix II.

Composition of the Four Prairie Soils, showing the Portion soluble in Acids and the Portion insoluble.

	No. I.	No. II.	No. III.	No. IV.
Soluble.				
*Organic matter and water of Combination	7·54	5·76	9·77	9·05
Alumina	2·80	1·57	3·51	3·38
Oxides of iron	4·95	2·57	4·13	4·30
Lime	·44	·35	·77	·54
Magnesia	·45	·40	·21	·35
Potash	·65	·34	·15	·19
Soda	trace.	trace.	·05	·08
Phosphoric acid	·08	·05	·12	·10
Sulphuric acid	·07	·05	·14	·08
Carbonic acid, traces of chlorine and loss	·74	·53	·82	·09
Insoluble.				
Insoluble silicates and land	(82·28)	(88·38)	(80·33)	(81·84)
Consisting of—				
Alumina	3·87	4·98	5·07	5·36
Lime	·93	none.	1·07	·59
Magnesia	·58	1·13	·61	·26
Potash	1·04	1·06	1·05	1·10
Soda	·82	·53	·78	·42
Silica	75·04	80·68	71·75	74·11
	100·00	100·00	100·00	100·00
* Containing Nitrogen	·30		·33	·34
Equal to Ammonia	·36	·26	·40	·41
		.31		

Augustus Voelcker.

Royal Agricultural College, Cirencester,
December 28*th*, 1858.

Appendix III.

"Royal Agricultural College,
Cirencester, Dec. 28th, 1858.

"Dear Sir,—I have now the pleasure of handing you the results of detailed and careful analyses of the four soils you sent to me about a month ago. At the same time I enclose copies of two remarkably fertile soils resting on the old red sandstone formation, as analysed by myself some time ago, and also a series of soil analyses made some years ago by Professor Anderson at request of the Highland Agricultural Society of Scotland.

"You will not fail to recognise a general similarity in the composition of your soils, and observe that they all contain large quantities of potash, no doubt in the state of easily decomposable silicate of potash. With the exception of No. 2, they also contain silicate of lime, which in No. 2 appears to be entirely replaced by silicate of magnesia.

"The soils are not very rich in phosphoric acid, but still there is amply sufficient to meet all the requirements of plants usually cultivated on the farm. However, I cannot help thinking that phosphatic manures, especially for root crops, should be used liberally on your soils, and guano and other ammoniacal manures more sparingly, for you will perceive that all four soils are very rich in nitrogenised organic matter. *Indeed, I have never analysed before soils which contained so much nitrogen, nor do I find any records of soils richer in nitrogen than yours.*

"In No. 2 the proportion of nitrogen is smaller than in the three other soils, which might have been expected, since there is not so much organic matter.

"In the soil from the Carse of Gowrie Dr. Anderson found 2 per cent. of nitrogen, which he considers a large proportion. Dr. Anderson, referring to this constituent, says, 'The actually large amount of this quantity may not be apparent when it is expressed in fractions of a per cent.; we must bear in mind that the weight of the soil ten inches deep on an acre is, in round numbers, about a thousand tons, and that quantity will contain about two tons of nitrogen. It will further serve to illustrate its abundance to mention that a crop of wheat, amounting to 36 bushels, with straw, contains about 52 lbs. of nitrogen, and a crop of Swedish turnips only about 36 lbs.

"In the least fertile of your four soils, and I consider No. 2 the least fertile, in a purely chemical point of view, there is more nitrogen than in the Carse soil. It is this large amount of nitrogen, and the beautiful state of division, that impart a peculiar character to your soils, and distinguish them so favourably.

"They are soils upon which I imagine flax could be grown in perfection, supposing the climate to be otherwise favourable. There is one particular to which I would direct your attention. None of these four soils contains, in comparison with other soils, a high percentage of lime, and No. 2 especially I think would be benefited by an occasional application of lime. The soil No. 3 contains most lime, both as carbonate and in the state of silicate of lime.

"I have analysed separately the portion soluble in acids, and the portion of the solids insoluble in acids. The analysis of the insoluble silicates requires fusion and takes up much time, and for this reason I could not send you the results before.

"For comparison's sake with other soils, I have put together the results obtained in the analysis of the portion soluble in acids, and those obtained in the analysis of the insoluble silicates, so that you may see at a glance the total amount of lime, potash, &c., in 100 parts of dry soil. In the same table I have given approximate determinations of the amount of sand, clay, lime and organic matter, which may be useful for some purposes.

"Believe me, dear Sir,

"Yours faithfully,

"AUGUSTUS VOELCKER.

"J. Caird, Esq., M. P."

THE END

Meta Gray; or, What Makes Home Happy. By MARIA J. McINTOSH, Author of "Aunt Kitty's Tales." 1 vol., 12mo. 75 cents.

The Emancipation of Faith. By the late HENRY EDWARD SCHEDEL, M. D. Edited by GEORGE SCHEDEL. 2 vols., 8vo. Cloth, $4.

The Ministry of Life. By MARIA LOUISA CHARLESWORTH, Author of "Ministering Children.
1 volume, 12mo. Cloth, with 2 engravings, $1.

Bertram Noel: A Story of Youth. By E. J. MAY, Author of "Edgar Clifton." 1 volume, 16mo. Illustrated, 75 cents.

Benton's Thirty Years' View; or, A History of the Working of the American Government for Thirty Years, from 1820 to 1850. New edition, with Autobiography and a General Index. 2 volumes, 8vo. Cloth, $5.

The Household Book of Poetry. Collected and Edited by CHARLES A. DANA. Third Edition. 1 volume, half morocco, $3 50.

New York to Delhi, by the way of Rio de Janeiro, Australia, and China. By ROBERT B. MINTURN, JR.
1 volume, 12mo. Illustrated with a Map, $1 25. (Second edt.)

History of Civilization in England. By HENRY THOMAS BUCKLE. Vol. 1, 8vo. 677 pages. From the 2d London edt., $2 50.

Rational Cosmology; or, The Eternal Principles and the Necessary Laws of the Universe. By LAURENS P. HICKOK, D. D. 1 volume, 8vo. 397 pages, $1 75.

Whewell's History of the Inductive Sciences. First American, from the third London edition. 2 vols, 8vo. Cloth, $4.

The Coopers; or, Getting Under Way. By ALICE B. HAVEN. 1 volume, 12mo. 336 pages, 75 cents.

Appleton's New American Cyclopædia. A Popular Dictionary of General Knowledge. Volume V. Just published. To be completed in fifteen volumes.
Cloth, $3; leather, $3 50; hf. mor. $4; hf. Russia, $4 50. Published by subscription.

Benton's Abridgment of the Congressional Debates. Volume X. Just published. Sold by Subscription.
Cloth, $3; law sheep, $3 50; half morocco, $4. Each volume payable as delivered.

Burton's Cyclopædia of Wit and Humor. Two large volumes, 8vo. Profusely illustrated with Wood Engravings and twenty-four Portraits on Steel.
Extra cloth, $7; sheep extra, $8; hf mor. $9; hf calf, $10.

www.ingramcontent.com/pod-product-compliance
Lightning Source LLC
LaVergne TN
LVHW022059200926
848481LV00003B/996